ELEMENTARY GEOGRAPHY & CULTURES

Parent Lesson Planner
(PLP)

 Weekly Lesson Schedule

Student Worksheets

 Quizzes

 Answer Key

3rd — 6th grade

| 1 Year Geography | |

First printing: March 2014

Second printing: June 2014

ISBN: 978-0-89051-808-3

Printed in the United States of America

Please visit our website for other great titles:
www.masterbooks.net

For information regarding author interviews,
please contact the publicity department at (870) 438-5288.

Master Books®
A Division of New Leaf Publishing Group
www.masterbooks.net

Where Creation Inspires Education

Since 1975, Master Books has been providing educational resources based on a biblical worldview to students of all ages. At the heart of these resources is our firm belief in a literal six-day creation, a young earth, the global Flood as revealed in Genesis 1–11, and other vital evidence to help build a critical foundation of scriptural authority for everyone. By equipping students with biblical truths and their key connection to the world of science and history, it is our hope they will be able to defend their faith in a skeptical, fallen world.

If the foundations are destroyed, what can the righteous do?
Psalm 11:3; NKJV

As the largest publisher of creation science materials in the world, Master Books is honored to partner with our authors and educators, including:

Ken Ham of Answers in Genesis

Dr. John Morris and Dr. Jason Lisle of the Institute for Creation Research

Dr. Donald DeYoung and Michael Oard of the Creation Research Society

Dr. James Stobaugh, John Hudson Tiner, Rick and Marilyn Boyer, Dr. Tom DeRosa, Todd Friel, Israel Wayne, and so many more!

Whether a pre-school learner or a scholar seeking an advanced degree, we offer a wonderful selection of award-winning resources for all ages and educational levels.

But sanctify the Lord God in your hearts, and always be ready
to give a defense to everyone who asks you a reason for the hope
that is in you, with meekness and fear.
1 Peter 3:15; NKJV

Permission to Copy

Lessons for a 36-week course!

Overview: This *Elementary Geography & Cultures PLP* contains materials for use with *The Children's Atlas of God's World* and *Passport to the World*. Materials are organized by each book in the following sections:

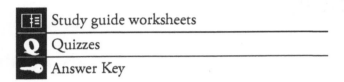

	Study guide worksheets
Q	Quizzes
	Answer Key

Features: Each suggested weekly schedule has four to five easy-to-manage lessons that combine reading, worksheets, and vocabulary-building opportunities. Worksheets and the quiz in this PLP are perforated and three-hole punched. Materials are easy to tear out, hand out, grade, and store. As always, you are encouraged to adjust the schedule and materials as you need to in order to best work within your educational program.

Workflow: Students will read the pages in their book and then complete each section of the course materials. Tests are given at regular intervals with space to record each grade. Younger students may be given the option of taking open-book tests.

Lesson Scheduling: Space is given for assignment dates. There is flexibility in scheduling. For example, the parent may opt for a M-W schedule rather than a M, W, F schedule. Each week listed has five days, but due to vacations the school work week may not be M-F. Please adapt the days to your school schedule. As the student completes each assignment, he/she should put an "X" in the box.

🕐	Approximately 30 to 45 minutes per lesson, four to five days a week
🔑	Includes answer keys for worksheets and quizzes
📑	Worksheets for every lesson
🔄	Quizzes are included to help reinforce learning and provide assessment opportunities
📄	Designed for grades 3 to 6 in a one-year course
⚗️	Supports activity-based learning

Craig Froman is a curriculum writer, as well as assistant editor at New Leaf Publishing Group, and a homeschooling father. He has a bachelor of arts in business administration from Vanguard University in Costa Mesa, California, and a master's degree in education from the Assemblies of God Theological Seminary in Springfield, Missouri.

The Children's Atlas of God's World - This atlas is packed with unique insights into Christian history makers and amazing landmarks. One will explore the design of ecosystems and biomes, great civilizations, and discoveries found around the world.

Passport to the World - Children will gain knowledge of intriguing facts about countries, including their capital cities, maps, flags, populations, and religions. Gathering stickers for their passport, readers learn greetings from 26 different languages, as well as interesting language tips.

Contents

Course Description

Children will travel the world in the comfort of their own homes. Through this *Elementary Geography & Cultures* course, students experience a wondrous global journey within the pages of two God-honoring books, encountering places and people from around the world. These full-color, visually engaging books provide a dual purpose as an elementary curriculum and as valued reference tools.

Passport to the World

Children will gain knowledge of intriguing facts about countries, including their capital cities, maps, flags, populations, and religions. Gathering stickers for their passport, readers learn greetings from 26 different languages, as well as interesting language tips. Discover various cultures and customs, and fill up the passport with stickers from the countries visited. You'll learn that:

- The language journey began just over 4,000 years ago at the Tower of Babel.

- There is a huge slab of limestone in Bolivia that has some 5,000 dinosaur footprints.

- A traditional Christmas Eve dinner in Lithuania includes 12 dishes, one for each of the Apostles.

- All Bengali literature was rhymed verse if written before the 19th century.

Passport to the World creates a fun and fact-filled adventure students can share with others through interactive games included in the back of the book and in the passport provided.

Children's Atlas of God's World

The atlases that line the shelves of libraries and bookstores are filled with both evolutionary thought and secular worldviews. This atlas is packed with unique insights into Christian history makers and amazing landmarks. One will explore the design of ecosystems and biomes, great civilizations, and discoveries found around the world. This atlas glorifies God, explores His creation, and honors His followers around the world! You'll learn about:

- Amazing wonders of God's creation, including longest rivers, tallest mountains, and more.

- Interesting factual details concerning Christian explorers, missionaries, and history makers.

- Geographic features and how these were formed by the Flood, plus other details of God's amazing design.

Outline maps and facts regarding the seven continents are provided, as well as detailed maps and data of the featured countries. The comprehensive information provided for each focus country will bring to light their culture and traditions, holidays, exploration, legal system, and economic industries, as well as missionary accounts and other material to help children connect to people from regions around the globe.

First Semester Suggested Daily Schedule

Date	Day	Assignment	Due Date	✓	Grade
		First Semester-First Quarter			
Week 1	Day 1	Read Pages 6-7 • History of Languages • *Passport to the World* • (PW) Read Pages 8-9 • Armenia • (PW)			
	Day 2	Armenia Worksheet • Page 15 • Lesson Planner • (LP)			
	Day 3	Armenia Activity • Page 16 • (LP)			
	Day 4	Read Pages 10-11 • Bangladesh • (PW)			
	Day 5	Bangladesh Worksheet • Page 17 • (LP)			
Week 2	Day 6	Bangladesh Activity • Page 18 • (LP)			
	Day 7	Read Pages 12-13 • United States of America • (PW)			
	Day 8	United States of America Worksheet • Page 19 • (LP)			
	Day 9	United States of America Activity • Page 20 • (LP)			
	Day 10	Read Pages 14-15 • Netherlands • (PW)			
Week 3	Day 11	Netherlands Worksheet • Page 21 • (LP)			
	Day 12	Netherlands Activity • Page 22 • (LP)			
	Day 13	Read Pages 16-17 • Australia • (PW)			
	Day 14	Australia Worksheet • Page 23 • (LP)			
	Day 15	Australia Activity • Page 24 • (LP)			
Week 4	Day 16	Read Pages 18-19 • France • (PW)			
	Day 17	France Worksheet • Page 25 • (LP)			
	Day 18	France Activity • Page 26 • (LP)			
	Day 19	Read Pages 20-21 • Germany • (PW)			
	Day 20	Germany Worksheet • Page 27 • (LP)			
Week 5	Day 21	Germany Activity • Page 28 • (LP)			
	Day 22	Read Pages 22-23 • Israel • (PW)			
	Day 23	Israel Worksheet • Page 29 • (LP)			
	Day 24	Israel Activity • Page 30 • (LP)			
	Day 25	Read Pages 24-25 • Iceland • (PW)			
Week 6	Day 26	Iceland Worksheet • Page 31 • (LP)			
	Day 27	Iceland Activity • Page 32 • (LP)			
	Day 28	Read Pages 26-27 • Japan • (PW)			
	Day 29	Japan Worksheet • Page 33 • (LP)			
	Day 30	Japan Activity • Page 34 • (LP)			
Week 7	Day 31	Read Pages 28-29 • South Korea • (PW)			
	Day 32	South Korea Worksheet • Page 35 • (LP)			
	Day 33	South Korea Activity • Page 36 • (LP)			
	Day 34	Read Pages 30-31 • Lithuania • (PW)			
	Day 35	Lithuania Worksheet • Page 37 • (LP)			

Date	Day	Assignment	Due Date	✓	Grade
	Day 36	Lithuania Activity • Page 38 • (LP)			
	Day 37	Read Pages 32-33 • China • (PW)			
Week 8	Day 38	China Worksheet • Page 39 • (LP)			
	Day 39	China Activity • Page 40 • (LP)			
	Day 40	Read Pages 34-35 • Norway • (PW)			
	Day 41	Norway Worksheet • Page 41 • (LP)			
	Day 42	Norway Activity • Page 42• (LP)			
Week 9	Day 43	Read Pages 36-37 • India • (PW)			
	Day 44	India Worksheet • Page 43 • (LP)			
	Day 45	India Activity • Page 44 • (LP)			
First Semester-Second Quarter					
	Day 46	Read Pages 38-39 • Afghanistan • (PW)			
	Day 47	Afghanistan Worksheet • Page 45 • (LP)			
Week 1	Day 48	Afghanistan Activity • Page 46 • (LP)			
	Day 49	Read Pages 40-41 • Bolivia • (PW)			
	Day 50				
	Day 51	Bolivia Worksheet • Page 47 • (LP)			
	Day 52	Bolivia Activity • Page 48 • (LP)			
Week 2	Day 53	Read Pages 42-43 • Russia • (PW)			
	Day 54	Russia Worksheet • Page 49 • (LP)			
	Day 55				
	Day 56	Russia Activity • Page 50 • (LP)			
	Day 57	Read Pages 44-45 • Mexico • (PW)			
Week 3	Day 58	Mexico Worksheet • Page 51 • (LP)			
	Day 59	Mexico Activity • Page 52 • (LP)			
	Day 60				
	Day 61	Read Pages 46-47 • Turkey • (PW)			
	Day 62	Turkey Worksheet • Page 53 • (LP)			
Week 4	Day 63	Turkey Activity • Page 54 • (LP)			
	Day 64	Read Pages 48-49 • Ukraine • (PW)			
	Day 65				
	Day 66	Ukraine Worksheet • Page 55 • (LP)			
	Day 67	Ukraine Activity • Page 56 • (LP)			
Week 5	Day 68	Read Pages 50-51 • Vietnam • (PW)			
	Day 69	Vietnam Worksheet • Page 57 • (LP)			
	Day 70				
	Day 71	Vietnam Activity • Page 58 • (LP)			
	Day 72	Read Pages 52-53 • United Kingdom • (PW)			
Week 6	Day 73	United Kingdom Worksheet • Page 59 • (LP)			
	Day 74	United Kingdom Activity • Page 60 • (LP)			
	Day 75				

Date	Day	Assignment	Due Date	✓	Grade
	Day 76	Read Pages 54-55 • South Africa • (PW)			
	Day 77	South Africa Worksheet • Page 61 • (LP)			
Week 7	Day 78	South Africa Activity • Page 62 • (LP)			
	Day 79	Read Pages 56-57 • Nigeria • (PW)			
	Day 80				
	Day 81	Nigeria Worksheet • Page 63 • (LP)			
	Day 82	Nigeria Activity • Page 64 • (LP)			
Week 8	Day 83	Read Pages 58-59 • Swaziland • (PW)			
	Day 84	Swaziland Worksheet • Page 65 • (LP)			
	Day 85				
	Day 86	Swaziland Activity • Page 66 • (LP)			
	Day 87	Read Pages 60-61 • Country Facts at a Glance • (PW)			
Week 9	Day 88	**Quiz** • Name That Flag • Page 62 • (PW)			
	Day 89	**Quiz** • Name That Country • Page 63 • (PW)			
	Day 90				
		Mid-Term Grade			

Second Semester Suggested Daily Schedule

Date	Day	Assignment	Due Date	✓	Grade
		Second Semester-Third Quarter			
Week 1	Day 91	Read Pages 2-3 • Intro • *Children's Atlas of God's World* • (CAGW)			
	Day 92	Read Page 6 • North America • (CAGW)			
	Day 93	Read Pages 8-9 • United States • (CAGW)			
	Day 94	Read Pages 10-11 • United States • (CAGW)			
	Day 95	United States Worksheet • Page 69 • (LP)			
Week 2	Day 96	United States Activity • Page 70 • (LP)			
	Day 97	Read Pages 12-13 • Mexico • (CAGW)			
	Day 98	Read Pages 14-15 • Mexico • (CAGW)			
	Day 99	Mexico Worksheet • Page 71 • (LP)			
	Day 100	Mexico Activity • Page 72 • (LP)			
Week 3	Day 101	Read Pages 16-17 • Canada • (CAGW)			
	Day 102	Read Pages 18-19 • Canada • (CAGW)			
	Day 103	Canada Worksheet • Page 73 • (LP)			
	Day 104	Canada Activity • Page 74 • (LP)			
	Day 105	Read Pages 20-21 • South America • (CAGW)			
Week 4	Day 106	Read Pages 22-23 • Brazil • (CAGW)			
	Day 107	Read Pages 24-25 • Brazil • (CAGW)			
	Day 108	Brazil Worksheet • Page 75 • (LP)			
	Day 109	Brazil Activity • Page 76 • (LP)			
	Day 110	Read Pages 26-27 • Bolivia • (CAGW)			
Week 5	Day 111	Bolivia Worksheet • Page 77 • (LP)			
	Day 112	Bolivia Activity • Page 78 • (LP)			
	Day 113	Read Pages 28-29 • Europe • (CAGW)			
	Day 114	Read Pages 30-31 • Norway • (CAGW)			
	Day 115	Norway Worksheet • Page 79 • (LP)			
Week 6	Day 116	Norway Activity • Page 80 • (LP)			
	Day 117	Read Pages 32-33 • Italy • (CAGW)			
	Day 118	Italy Worksheet • Page 81 • (LP)			
	Day 119	Italy Activity • Page 82 • (LP)			
	Day 120	Read Pages 34-35 • France • (CAGW)			
Week 7	Day 121	Read Pages 36-37 • France • (CAGW)			
	Day 122	France Worksheet • Page 83 • (LP)			
	Day 123	France Activity • Page 84 • (LP)			
	Day 124	Read Pages 38-41 • Germany • (CAGW)			
	Day 125	Germany Worksheet • Page 85 • (LP)			

Date	Day	Assignment	Due Date	✓	Grade
	Day 126	Germany Activity • Page 86 • (LP)			
	Day 127	Read Pages 42-45 • United Kingdom • (CAGW)			
Week 8	Day 128	United Kingdom Worksheet • Page 87 • (LP)			
	Day 129	United Kingdom Activity • Page 88 • (LP)			
	Day 130	Read Pages 46-47 • Russia • (CAGW)			
	Day 131	Read Pages 48-49 • Russia • (CAGW)			
	Day 132	Russia Worksheet • Page 89 • (LP)			
Week 9	Day 133	Russia Activity • Page 90 • (LP)			
	Day 134	Read Pages 50-51 • Africa • (CAGW)			
	Day 135	Read Pages 52-53 • Kenya • (CAGW)			
Second Semester-Fourth Quarter					
	Day 136	Kenya Worksheet • Page 91 • (LP)			
	Day 137	Kenya Activity • Page 92 • (LP)			
Week 1	Day 138	Read Pages 54-55 • Egypt • (CAGW)			
	Day 139	Egypt Worksheet • Page 93 • (LP)			
	Day 140	Egypt Activity • Page 94 • (LP)			
	Day 141	Read Pages 56-57 • South Africa • (CAGW)			
	Day 142	South Africa Worksheet • Page 95 • (LP)			
Week 2	Day 143	South Africa Activity • Page 96 • (LP)			
	Day 144	Read Pages 58-59 • Asia • (CAGW)			
	Day 145	Read Pages 60-61 • Israel • (CAGW)			
	Day 146	Read Pages 62-63 • Israel • (CAGW)			
	Day 147	Israel Worksheet • Page 97 • (LP)			
Week 3	Day 148	Israel Activity • Page 98 • (LP)			
	Day 149	Read Pages 64-65 • Saudi Arabia • (CAGW)			
	Day 150	Saudi Arabia Worksheet • Page 99 • (LP)			
	Day 151	Saudi Arabia Activity • Page 100 • (LP)			
	Day 152	Read Pages 66-67 • India • (CAGW)			
Week 4	Day 153	Read Pages 68-69 • India • (CAGW)			
	Day 154	India Worksheet • Page 101 • (LP)			
	Day 155	India Activity • Page 102 • (LP)			
	Day 156	Read Pages 70-71 • China • (CAGW)			
	Day 157	Read Pages 72-73 • China • (CAGW)			
Week 5	Day 158	China Worksheet • Page 103 • (LP)			
	Day 159	China Activity • Page 104 • (LP)			
	Day 160	Read Pages 74-75 • Japan • (CAGW)			

Date	Day	Assignment	Due Date	✓	Grade
Week 6	Day 161	Read Pages 76-77 • Japan • (CAGW)			
	Day 162	Japan Worksheet • Page 105 • (LP)			
	Day 163	Japan Activity • Page 106 • (LP)			
	Day 164	Read Pages 78-79 • Malaysia • (CAGW)			
	Day 165	Malaysia Worksheet • Page 107 • (LP)			
Week 7	Day 166	Malaysia Activity • Page 108 • (LP)			
	Day 167	Read Pages 80-81 • Australia/Oceania • (CAGW)			
	Day 168	Read Pages 82-83 • Australia • (CAGW)			
	Day 169	Australia Worksheet • Page 109 • (LP)			
	Day 170	Australia Activity • Page 110 • (LP)			
Week 8	Day 171	Read Pages 84-85 • New Zealand • (CAGW)			
	Day 172	New Zealand Worksheet • Page 111 • (LP)			
	Day 173	New Zealand Activity • Page 112 • (LP)			
	Day 174	Read Pages 86-87 • Antarctica • (CAGW)			
	Day 175				
Week 9	Day 176	Antarctica Worksheet • Page 113 • (LP)			
	Day 177				
	Day 178	Read Pages 88-89 • Biomes of the World • (CAGW)			
	Day 179	**Quiz** • Biomes of the World • Page 117 • (LP)			
	Day 180				
		Final Grade			

Geography Worksheets

for Use with

Passport to the World

Short Answers

1. What is the capital city of Armenia?

2. How do you say "hello" in Armenian?

3. How do you say "goodbye" in Armenian?

4. How do you say "thank you" in Armenian?

5. How do you say "peace" in Armenian?

6. Approximately how many people in the world speak Armenian?

7. Finish the proverb: As mills require two stones, so friendship requires _____ _____.

8. Noah's ark came to rest on the mountain range of _____ in Armenia.

9. A restaurant in Armenia that serves sandwiches and coffee is called a _____.

10. The Armenian language borrowed Greek, Persian, Russian, and _____ words.

Color in the flag of Armenia

```
┌─────────────────────────┐
│                         │
├─────────────────────────┤
│                         │
├─────────────────────────┤
│                         │
└─────────────────────────┘
```

Make some gata with your parent/teacher.

Dough

1 tablespoon dry yeast
1 cup sour cream
1 cup unsalted butter
1 egg
1 tablespoon vegetable oil
1 tablespoon vinegar
3 cups sifted all-purpose flour,
plus ½ cup flour, for kneading

Filling

1 cup butter, melted
2 cups sifted all-purpose flour
1¼ cups sugar
½ teaspoon vanilla

Glaze

2 egg yolks, beaten
1 teaspoon yogurt

Directions:

1. In a mixing bowl, combine the yeast with 1 cup sour cream; set aside for 10 minutes. Add butter and mix well. Add the egg, oil, and vinegar and mix well. Add the sifted flour gradually and continue mixing. Knead dough for about 15 minutes until smooth and firm so that it does not stick to your hands. Gather dough into a ball and cover with plastic wrap. Refrigerate the dough overnight. To make the filling, mix together 1 cup of the melted butter and 2 cups flour. Add 1¼ cups sugar and ½ teaspoon vanilla; stir constantly for 1 minute, so that the mixture does not stick to your hands and becomes smooth and even. Set aside.

2. Preheat oven to 350°F. Remove dough from the refrigerator; divide it into 8 equal balls. Place each ball on a lightly floured board. Roll each ball out to a rectangle measuring 10 by 6 inches, or as thin as possible with the aid of a rolling pin. Paint each rectangle with melted butter. Spread ¼ cup filling into the center of each rectangle and roll out the filling over it with the aid of a rolling pin. Fold in about ½ inch of the rectangle on each side. Roll up dough into a cylinder. Place rolling pin on pastry roll and flatten slightly lengthwise. Cut the roll diagonally with a sharp knife or serrated knife into 2-inch equal slices. Place pastry about 2 inches apart on a lightly floured baking pan. (Be careful not to spill out any filling.)

3. Paint the surface of each pastry with the glaze (mixture of beaten egg). Place baking pan in the oven and bake for 15 to 30 minutes, or until golden brown. Remove the baking pan from oven and allow the pastries to cool.

From food.com

Short Answers

1. What is the capital city of Bangladesh?

2. How do you say "hello" in Bengali?

3. How do you say "goodbye" in Bengali?

4. How do you say "thank you" in Bengali?

5. How do you say "peace" in Bengali?

6. Approximately how many people in the world speak Bengali?

7. Finish the proverb: A handful of love is better than an oven full of _____.

8. Bengali literature was in _____ verse if written prior to the 19th century.

9. The Sundarbans National Park contains a coastal _____ forest.

10. As much as _____ percent of the land is at times covered in water.

Activities

Color in the flag of Bangladesh

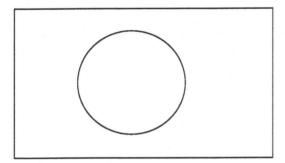

Make one of your favorite stir-fry veggie, chicken, or steak meals with your parent/teacher, and add some ginger to help flavor it!

Short Answers

1. What is the capital city of the United States?

2. How do you say "hello" in Cherokee?

3. How do you say "goodbye" in Cherokee?

4. How do you say "thank you" in Cherokee?

5. How do you say "peace" in Cherokee?

6. Approximately how many people in the world speak Cherokee?

7. Finish the proverb: Always remember that a smile is something sacred, to be _____.

8. There are approximately _____ million takeoffs per year at American airports.

9. Most speakers of Cherokee live in either _____ or _____.

10. The postal service in the United States moves over _____ percent of the world's mail each day.

Activities

Color in the flag of the United States.

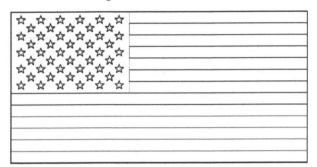

Make some pizza with your parent/teacher. You can get premade pizza from the store, or completely make your own. If you get a plain cheese pizza from the store, you can add all your own favorite toppings . . . peppers, onions, pepperoni, olives, mushrooms, and more! Most pizzas will be done in the oven within 10 to 15 minutes. Enjoy!

Short Answers

1. What is the capital city of the Netherlands?

2. How do you say "hello" in Dutch?

3. How do you say "goodbye" in Dutch?

4. How do you say "thank you" in Dutch?

5. How do you say "peace" in Dutch?

6. Approximately how many people in the world speak Dutch?

7. Finish the proverb: A _____ at one's back is a safe bridge.

8. One-_____ of the Netherlands is actually below sea level.

9. Wooden _____ were worn to protect workers in factories and mines.

10. Many people enjoy eating raw _____ with onions sprinkled over it.

Activities

Color in the flag of the Netherlands.

```
┌─────────────────────────┐
│                         │
├─────────────────────────┤
│                         │
├─────────────────────────┤
│                         │
└─────────────────────────┘
```

Make some herring and onion with your parent/teacher. Often this is simply fresh raw herring that is served up with onions and maybe pickle slices as well. You might want to prepare and cook the herring or other fish (trout, salmon, etc.) as you normally would with the onions, then add sliced pickles if you like at the end.

Short Answers

1. What is the capital city of Australia?

2. What is the typical greeting in English?

3. How do you say farewell in English?

4. What words do you use to show you are grateful in English?

5. What word to you use to express serenity in English?

6. Approximately how many people in the world speak English?

7. Finish the proverb: You can't judge a _____ by its cover.

8. In Australia there are more than _____ square miles of land for every person.

9. Nearly _____ of the world speaks English as either a first or second language.

10. The _____ is a unique wind instrument to Australia, and is thought to be over a thousand years old.

Color in the flag of Australia.

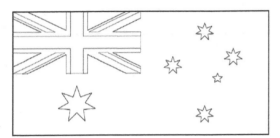

Make some toast with vegemite with your parent/teacher. You can buy vegemite at some grocery and specialty stores. Try a little first before spreading too much, to see if you like it!

Short Answers

1. What is the capital city of France?

2. How do you say "hello" in French?

3. How do you say "goodbye" in French?

4. How do you say "thank you" in French?

5. How do you say "peace" in French?

6. Approximately how many people in the world speak French?

7. Finish the proverb: Hunger is the best _____.

8. The _____ Tower always tilts away from the sun because the heat makes the metal expand.

9. France has over _____ kinds of cheese.

10. French was actually the official language of _____ for many years.

Activities

Color in the flag of France.

Try a variety of cheeses with your parent/teacher, like bleu, brie, camembert, Gouda, Munster, or Swiss.

Short Answers

1. What is the capital city of Germany?

2. How do you say "hello" in German?

3. How do you say "goodbye" in German?

4. How do you say "thank you" in German?

5. How do you say "peace" in German?

6. Approximately how many people speak German in Germany and Austria?

7. Finish the proverb: Better an _____ enemy than a false friend.

8. The tallest church in the world is the _____ Cathedral.

9. The first Bible ever printed was printed in _____ in 1456.

10. Baked _____ have been a part of German foods since at least the 12th century.

Activities

Color in the flag of Germany.

```
┌─────────────────────────┐
│                         │
├─────────────────────────┤
│                         │
├─────────────────────────┤
│                         │
└─────────────────────────┘
```

Make some pretzels with your parent/teacher.

Ingredients

1½ cups water	3 cups flour
1¼ teaspoons active dry yeast (1 pkg.)	2 cups water
2 tablespoons brown sugar	2 tablespoons baking soda
1¼ teaspoons salt	coarse salt
1 cup bread flour	2–4 tablespoons butter (melted)

Directions

1. Sprinkle yeast on lukewarm water in mixing bowl; stir to dissolve.
2. Add sugar and salt and stir to dissolve; add flour and knead dough until smooth and elastic.
3. Let rise at least ½ hour.
4. While dough is rising, prepare a baking soda water bath with 2 cups warm water and 2 tbsp. baking soda.
5. Be certain to stir often.
6. After dough has risen, pinch off bits of dough and roll into a long rope (about ½ inch or less thick) and shape.
7. Dip pretzel in soda solution and place on greased baking sheet.
8. Allow pretzels to rise again.
9. Bake in oven at 450° for about 10 minutes or until golden.
10. Brush with melted butter.
11. Toppings: After you brush with butter try sprinkling with coarse salt.

From food.com

Short Answers

1. What is the capital city of Israel?

2. How do you say "hello" in Hebrew?

3. How do you say "goodbye" in Hebrew?

4. How do you say "thank you" in Hebrew?

5. How do you say "peace" in Hebrew?

6. Approximately how many people speak Hebrew in Israel?

7. Finish the proverb: Promise _____ and do much.

8. The lowest dry land on earth is found in Israel around the _____ Sea.

9. Hebrew is the original language of the _____ Testament.

10. The money in Israel has _____ printed into it.

Activities

Color in the flag of Israel.

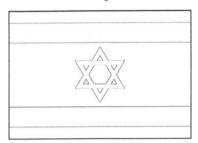

Make some potato latkes with your parent/teacher.

Ingredients

6 large potatoes, peeled

1 large onion

2 eggs, beaten

1 teaspoon baking powder

salt, to taste

pepper, to taste

1 pinch sugar

flour (enough to hold ingredients together, you'll be able to tell when you've added enough)

Crisco shortening

Directions

1. Cut peeled potatoes into chunks and chop in food processor.
2. Cut onions into chunks and chop in food processor.
3. Mix potatoes and onions together in a colander over a large bowl.
4. Allow to drain.
5. Pour mixture into a large bowl and add beaten eggs, salt, pepper, baking powder, and sugar. Add a tablespoon of flour at a time until mixture holds together.
6. Mix well.
7. In a fry pan, melt shortening.
8. You'll need about ⅛".
9. Spoon heaping tablespoons of the mixture into the oil.
10. Flatten each spoonful with the back of the spoon to make thin latkes.
11. Fry until the edges turn a dark brown.
12. Flip over to fry the other side.
13. Drain on a paper towel.
14. Serve with applesauce and/or sour cream.

From food.com

Short Answers

1. What is the capital city of Iceland?

2. How do you say "hello" in Icelandic?

3. How do you say "goodbye" in Icelandic?

4. How do you say "thank you" in Icelandic?

5. How do you say "peace" in Icelandic?

6. Approximately how many people speak Icelandic in Iceland?

7. Finish the proverb: Better shoeless than _____.

8. Iceland has around _____ volcanic mountains.

9. The small Icelandic _____ has hardly changed in over a thousand years.

10. The Icelandic _____ are delicate, papery flowers.

Activities

Color in the flag of Iceland.

Plant some poppies in a pot or garden with your parent/teacher. Take proper care to water them, and they can come back in a garden year after year.

Short Answers

1. What is the capital city of Japan?

2. How do you say "hello" in Japanese?

3. How do you say "goodbye" in Japanese?

4. How do you say "thank you" in Japanese?

5. How do you say "peace" in Japanese?

6. Approximately how many people speak Japanese in Japan?

7. Finish the proverb: Adversity is the foundation of _____.

8. They discovered ancient _____ structures in the ocean off the Japanese coast.

9. Skyscrapers are built to be able to sway in Japan because of all the _____.

10. _____ is one of the foods grown extensively across Japan.

Activities

Color in the flag of Japan.

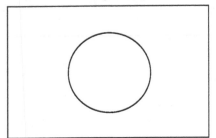

Make a meal with rice together with your parent/teacher. This might include stir-fry vegetables, as well as cut-up chicken or strips of steak. Soy sauce can add a distinctive Asian flavoring. Low-sodium sauces are available.

Short Answers

1. What is the capital city of South Korea?

2. How do you say "hello" in Korean?

3. How do you say "goodbye" in Korean?

4. How do you say "thank you" in Korean?

5. How do you say "peace" in Korean?

6. Approximately how many people speak Korean in South Korea?

7. Finish the proverb: Cast no _____ into the well that gives you water.

8. The founding of the Korean alphabet is celebrated on _____ 9th.

9. What is the monetary unit for South Korea?

10. _____ is a Korean food made from fermented vegetables.

Activities

Color in the flag of South Korea.

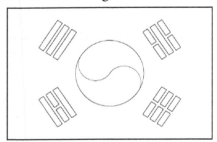

Make some kimchi with your parent/teacher.

Ingredients

14 cups napa cabbage, coarsley chopped (about 2 pounds)

3 tablespoons salt

1½ cups water

1 cup green onion, coarsley chopped

4 teaspoons fresh ginger, peeled and grated

1½ teaspoons red pepper flakes, crushed

4 garlic cloves, minced

Directions

1. Place cabbage and salt in a large bowl, tossing gently to combine.
2. Weigh down cabbage with another bowl.
3. Let stand at room temperature 3 hours, tossing occasionally.
4. Drain and rinse with cold water.
5. Drain and squeeze.
6. Combine cabbage, water, and remaining ingredients.
7. Cover and refrigerate at least 4 hours.

From food.com

Short Answers

1. What is the capital city of Lithuania?

2. How do you say "hello" in Lithuanian?

3. How do you say "goodbye" in Lithuanian?

4. How do you say "thank you" in Lithuanian?

5. How do you say "peace" in Lithuanian?

6. Approximately how many people speak Lithuanian in Lithuania?

7. Finish the proverb: Fear and _____ do not go together.

8. What kind of monetary unit is used in Lithuania?

9. Bread and _____ are considered very important in the culture.

10. In a traditional Christmas Eve dinner, _____ dishes are served.

Activities

Color in the flag of Lithuania.

```
┌─────────────────────────────────┐
│                                 │
│                                 │
├─────────────────────────────────┤
│                                 │
├─────────────────────────────────┤
│                                 │
│                                 │
└─────────────────────────────────┘
```

Make some cepelinai with your parent/teacher.

Ingredients

8 ounces farmers' cheese
6 tablespoons farina (cream of wheat)
3 tablespoons potato starch
1 egg
½ teaspoon salt
Additional potato starch

Filling

1 teaspoon butter or 1 teaspoon vegetable oil
½ small onion, finely chopped
3–4 ounces ground meat
1 pinch thyme or 1 pinch allspice
⅛ teaspoon salt
couple of grinds black pepper

Sauce

3 slices lean bacon
2 medium mushrooms, chopped
2–3 tablespoons butter

Directions

1. In a medium bowl, combine the cheese, potato starch, egg, and salt. I first stir with a spoon, but then knead it right in the bowl with my hand, to ensure that everything is thoroughly mixed. Cover and let it stand while you make the filling. (Letting it stand about 10 or 15 minutes lets the farina hydrate and I think makes the dough hold together better.)
2. Saute the onion in the oil in a small saucepan until translucent.
3. Add to the remaining filling ingredients in a small bowl. Mix thoroughly.
4. Divide into four portions and form into small sausage shapes. Cover.
5. Sprinkle two or three tablespoons of potato starch onto a large plate and set aside.
6. Fill a 2 or 3-quart saucepan with water, add salt, and bring to a boil while making the dumplings.
7. Divide the dough into four portions — I do this right in the bowl, just cutting through it with a knife.
8. Dampen your hands with cool water before forming each dumpling. Take one portion of the dough and flatten it on the palm of one hand into an oval.
9. Place one portion of the filling onto the dough and form the dough around it into an oval shape. It's important to make the exterior very smooth, without any cracks or seams — this will keep the dumplings from splitting as they cook.
10. Place the dumpling onto the potato-starch-sprinkled plate and roll it around to dredge in the starch.
11. Repeat with the remaining dough and filling.
12. When they are all done, carefully place each dumpling into the pot of boiling water. They should fit without crowding or overlapping. If there is any potato starch left on the plate, just scrape it right into the water, as well.
13. Let the water return to a boil and reduce it to a slow boil — more than a simmer, but not a full boil.
14. Cook for about 10 minutes after it returns to a boil.
15. While these cook, make the sauce.
16. Chop the bacon and saute it in a small saucepan.
17. As it starts to get translucent, add the mushrooms and butter, and saute until cooked through.
18. Remove the dumplings with a slotted spoon, place on a plate and cover with sauce.

From food.com

Short Answers

1. What is the capital city of China?

2. How do you say "hello" in Mandarin?

3. How do you say "goodbye" in Mandarin?

4. How do you say "thank you" in Mandarin?

5. How do you say "peace" in Mandarin?

6. Approximately how many people speak Mandarin in China?

7. Finish the proverb: A book holds a house of _____.

8. The Dragon's _____ are terraces begun some 700 years ago.

9. There are only around _____ family names in China.

10. What monetary unit does China use?

Activities

Color in the flag of China.

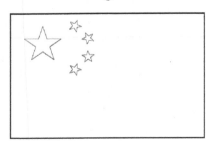

Make some Chinese ice cream with your parent/teacher. If you don't have any snow, you can blend up some ice, add some milk for texture, and just a few spoonfuls of cooked rice. Enjoy!

Short Answers

1. What is the capital city of Norway?

2. How do you say "hello" in Norwegian?

3. How do you say "goodbye" in Norwegian?

4. How do you say "thank you" in Norwegian?

5. How do you say "peace" in Norwegian?

6. Approximately how many people speak Norwegian in Norway?

7. Finish the proverb: On the road between the homes of friends, _____ does not grow.

8. What kind of monetary unit do they use in Norway?

9. Norway has around _____ glaciers.

10. _____ Ibsen wrote the play called *The Doll's House*.

Activities

Color in the flag of Norway.

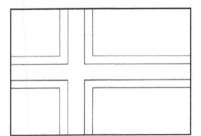

Make some lutefisk with your parent/teacher.

Ingredients

2–3 lbs lutefisk

2 teaspoons salt

½ cup melted butter

Directions

1. Thaw lutefisk if frozen and cut into serving pieces.
2. Rinse and drain well. Place fish skin side down in one layer of heavy foil.
3. Sprinkle with salt.
4. Bring foil around and make a fold in it.
5. Fold up ends and seal.
6. Place in shallow glass pan on a rack, seam up and bake at 325° for at least one hour.
7. I do it a little more.
8. Cut corner and drain out excess water.
9. Serve with melted butter on a heated platter.

From food.com

Short Answers

1. What is the capital city of India?

2. How do you say "hello" in Oriya?

3. How do you say "goodbye" in Oriya?

4. How do you say "thank you" in Oriya?

5. How do you say "peace" in Oriya?

6. Approximately how many people in the world speak Oriya?

7. Finish the proverb: If you live in the river you should make friends with the _____.

8. The river _____ inspired the name India.

9. There are more _____ in India than in any other country, the first created around 700 B.C.

10. In India, the _____ is the kind of money used.

Activities

Color in the flag of India.

Make a vegetable dish with curry with your parent/teacher.

Short Answers

1. What is the capital city of Afghanistan?

2. How do you say "hello" in Pashto?

3. How do you say "goodbye" in Pashto?

4. How do you say "thank you" in Pashto?

5. How do you say "peace" in Pashto?

6. Approximately how many people speak Pashto in Afghanistan?

7. Finish the proverb: Skill is stronger than _____.

8. _____ fighting and running are popular sports here.

9. The monetary unit used in Afghanistan is _____.

10. The Tajiks of Afghanistan speak _____.

Activities

Color in the flag of Afghanistan.

Make some naan with your parent/teacher.

Ingredients

4 cups flour
1 teaspoon baking powder
1 teaspoon salt
2 cups plain low-fat yogurt

Directions

1. Mix together flour, baking powder, and salt.

2. Stir in the yogurt until the dough is too stiff for a spoon, then knead it in the bowl until it holds together well, adding more flour if necessary.

3. Turn it out on a floured surface and continue kneading for about 5 minutes until the dough feels smooth and elastic.

4. Form the dough into a ball and put it in an oiled bowl, covered with a towel, to rest for an hour or longer.

5. Take the dough out and cut it into 10 equal pieces. Form each into a ball and press the balls flat into round discs.

6. Heat a large frying pan or griddle, either a seasoned cast iron or a good non-stick finish.

7. Heat your oven to about 500° and have the broiler on (this is how the original recipe states it — I know with my oven it's either 500° oven OR the broiler, but you get the idea).

8. Take 1 piece of dough at a time and roll it out on a floured surface till it is about 8–10 inches across and less than ¼ inch thick.

9. Lay it on the hot griddle and cook it over a medium heat for 4–5 minutes (I don't think mine took that long).

10. It will puff up in places or all over, and there will be some blackish-brown spots on the bottom.

11. Slide a spatula under the naan and transfer it to the oven, directly onto the rack, for a minute or two, just until it finishes puffing up into a balloon and begins to color lightly on top.

12. Remove naan from the oven and brush it lightly with melted butter if you like.

13. Continue this way with all the dough, stacking the breads into a napkin-lined basket.

14. Serve the breads hot, fresh from the oven, or let them cool and wrap them up.

15. To reheat, wrap them in aluminum foil, in packets of 4 or 5 breads and put them in a 400° oven for 10–15 minutes.

From food.com

Short Answers

1. What is the capital city of Bolivia?

2. How do you say "hello" in Quechua?

3. How do you say "goodbye" in Quechua?

4. How do you say "thank you" in Quechua?

5. How do you say "peace" in Quechua?

6. Approximately how many people speak Quechua in Bolivia?

7. Finish the proverb: Do not steal, do not lie, don't be _____.

8. An ancient _____ was found deep within Lake Titicaca.

9. The Quechua people are descendants of the _____.

10. The _____ mate plant is made into a drink much like tea.

Activities

Color in the flag of Bolivia.

Make some tea with your parent/teacher; herbal, green, black, or white!

Short Answers

1. What is the capital city of Russia?

2. How do you say "hello" in Russian?

3. How do you say "goodbye" in Russian?

4. How do you say "thank you" in Russian?

5. How do you say "peace" in Russian?

6. What kind of money do they use in Russia?

7. Finish the proverb: We do not care of what we have, but we _____ when it is lost.

8. Russia is the _____ country in the world.

9. It has long been a Russian custom to share tea and _____ cakes.

10. A train ride across Russia takes around a full _____.

Activities

Color in the flag of Russia.

```
┌─────────────────────────────┐
│                             │
│                             │
├─────────────────────────────┤
│                             │
│                             │
├─────────────────────────────┤
│                             │
│                             │
└─────────────────────────────┘
```

Make some tea and honey cakes with your parent/teacher.

Ingredients

1 cup honey, warmed in a pot over hot water

4 eggs, beaten

1½ cups flour

1 teaspoon baking powder

Directions

1. Preheat oven to 375°F, and butter and flour a 7" cake pan.

2. Beat honey until frothy. Add eggs, flour, and baking powder.

3. Pour into prepared cake pan. Bake for 15 minutes and check the cake. When done, it will shrink slightly from the pan. Continue baking for 5 minutes if it isn't ready yet (continue if it needs more time).

4. Turn it out of the pan while still hot, and let cool on a rack. Store in an airtight container for a day before serving.

From food.com

Short Answers

1. What kind of money do they use in Mexico?

2. How do you say "hello" in Spanish?

3. How do you say "goodbye" in Spanish?

4. How do you say "thank you" in Spanish?

5. How do you say "peace" in Spanish?

6. Approximately how many people speak Spanish in the world?

7. Finish the proverb: A wise man changes his _____, a fool never.

8. Mexico is the historic homeland of the ancient _____ Empire.

9. In Mexico, children take the _____ of both their mother and their father.

10. What is the largest city in the world?

Activities

Color in the flag of Mexico.

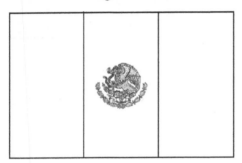

Make or heat up some corn tortillas with your parent/teacher.

Short Answers

1. What is the capital city of Turkey?

2. How do you say "hello" in Turkish?

3. How do you say "goodbye" in Turkish?

4. How do you say "thank you" in Turkish?

5. How do you say "peace" in Turkish?

6. What money system do they use in Turkey?

7. Finish the proverb: No road is long with _____ company.

8. Early Christians lived in the _____ formations found in the area of Cappadocia.

9. The only city in the world on two continents is _____.

10. Coffee houses in Turkey are called _____.

Activities

Color in the flag of Turkey.

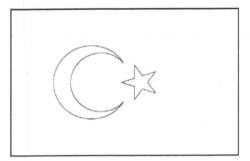

Make some halva with your parent/teacher.

Ingredients
½ cup unsalted butter
2 cups farina (the powder form, it should resemble fine cracked wheat or bulgur)
3 cups sugar
4 cups water

Directions
1. Melt the butter over a medium flame.
2. Once the butter begins to bubble, pour in the farina and toast to a light golden brown.
3. Add the sugar and mix thoroughly. Dont take too long; the more toasted the farina is, the more crumbly it will be.
4. Add the water and let it boil out. (Keep an eye on it because it will burn quickly if the water runs out.)
5. Let it cool and enjoy!

From food.com

Short Answers

1. What is the capital city of Ukraine?

2. How do you say "hello" in Ukrainian?

3. How do you say "goodbye" in Ukrainian?

4. How do you say "thank you" in Ukrainian?

5. How do you say "peace" in Ukrainian?

6. What kind of money do they use in Ukraine?

7. Finish the proverb: Only when you have eaten a _____ do you appreciate what sugar is.

8. Ukraine was once a part of the _____ _____.

9. Ukraine is often called the "_____ basket of Europe."

10. The _____ is a beautifully ornamented Ukrainian Easter egg.

Activities

Color in the flag of Ukraine.

Make some fresh bread with your parent/teacher. This can be made from a simple family recipe or from dough ready to bake from the store.

Short Answers

1. What is the capital city of Vietnam?

2. How do you say "hello" in Vietnamese?

3. How do you say "goodbye" in Vietnamese?

4. How do you say "thank you" in Vietnamese?

5. How do you say "peace" in Vietnamese?

6. What kind of money do they use in Vietnam?

7. Finish the proverb: A day of traveling will bring a basketful of _____.

8. Vietnam gained their freedom in A.D. _____.

9. _____ differs slightly depending on where you are in Vietnam.

10. French _____ planted the first coffee in 1857.

Activities

Color in the flag of Vietnam.

Make some stir-fried vegetables or a meat dish with your parent/teacher.

Short Answers

1. What is the capital city of the United Kingdom?

2. How do you say "hello" in Welsh?

3. How do you say "goodbye" in Welsh?

4. How do you say "thank you" in Welsh?

5. How do you say "peace" in Welsh?

6. Approximately how many people speak Welsh in the world?

7. Finish the proverb: A nation without a _____ is a nation without a heart.

8. The _____ Coast of Dorset, England, is considered a World Heritage Coastline.

9. People in the United Kingdom drink more _____ on average than any other country in the world.

10. _____ Castle happens to be the oldest royal home in the world that is still lived in.

Activities

Color in the flag of the United Kingdom.

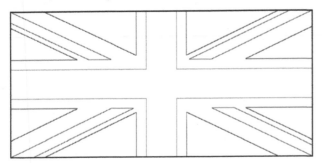

Make some Welsh Rabbit (Rarebit) with your parent/teacher.

Ingredients
6 slices toasted bread, buttered
1 teaspoon butter
½ lb extra-sharp cheddar cheese, shredded
1 teaspoon yellow mustard (or more)
1 dash hot sauce
1 cup apple juice
3 eggs, beaten
1 tomato, cut into sixths (wedges)

Directions
1. Melt butter in large saucepan; add cheese, mustard, and hot sauce; heat until cheese melts, stirring all the while.
2. Gradually add the apple juice, stirring continuously; then add meat if using.
3. Stir in beaten egg, until egg sets; remove from heat.
4. Cut toast into "points" and lay two points on each plate.
5. Pour one-sixth of the sauce over each plate; garnish with one tomato wedge.

From food.com

Short Answers

1. What is the legislative capital city of South Africa?

2. How do you say "hello" in Xhosa?

3. How do you say "goodbye" in Xhosa?

4. How do you say "thank you" in Xhosa?

5. How do you say "peace" in Xhosa?

6. Approximately how many people speak Xhosa in South Africa?

7. Finish the proverb: Throats are all alike in _____.

8. Nearly half of the world's _____ comes from South Africa.

9. What kind of money do they use in South Africa?

10. South Africa actually has _____ capital cities.

Activities

Color in the flag of South Africa.

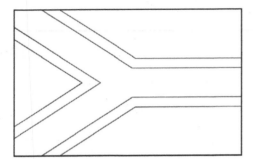

Try different types of dried meats with your parent/teacher. This might include beef, alligator, or ostrich, if you can find it!

Short Answers

1. What is the capital city of Nigeria?

2. How do you say "hello" in Yoruba?

3. How do you say "goodbye" in Yoruba?

4. How do you say "thank you" in Yoruba?

5. How do you say "peace" in Yoruba?

6. Approximately how many people speak Yoruba in Nigeria?

7. Finish the proverb: A proverb is the _____ that can carry one swiftly to the discovery of ideas.

8. What monetary unit is used in Nigeria?

9. Not far from the capital is a monolith called _____ Rock.

10. A game called _____ is played with a board of 12 cups and seeds.

Activities

Color in the flag of Nigeria.

```
+--------+--------+--------+
|        |        |        |
|        |        |        |
|        |        |        |
|        |        |        |
|        |        |        |
+--------+--------+--------+
```

Make some fried plantains with your parent/teacher. They can be sliced in ¼-inch sections and fried over medium heat in a pan with ¼ cup of water and butter or olive oil for flavor, if desired. Cook until slightly browned and crispy.

Short Answers

1. What is the capital city of Swaziland?

2. How do you say "hello" in Zulu?

3. How do you say "goodbye" in Zulu?

4. How do you say "thank you" in Zulu?

5. How do you say "peace" in Zulu?

6. What kind of money do they use in Swaziland?

7. Finish the proverb: Follow the _____ or flee the country.

8. Nguni people build their huts in the shape of _____.

9. The name Swaziland comes from a former king, King _____.

10. The first Zulu book of grammar actually came out in _____.

Activities

Color in the flag of Swaziland.

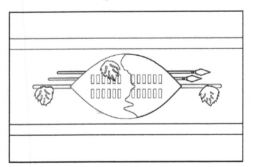

Enjoy some sliced avocado with your parent/teacher.

Geography Worksheets

for Use with

Children's Atlas of God's World

Short Answers

1. The deepest place in the ocean is the _____ Deep in the Mariana Trench.

2. Mount _____ is the tallest mountain on the North American continent.

3. The largest island in the world is _____.

4. What is the capital of the United States?

5. The national anthem of the United States is _____.

6. The Declaration of Independence states that people are "endowed by their _____ with certain unalienable rights."

7. A coniferous forest is one that has trees with cones and _____.

8. The Grand Canyon was carved by the retreating waters of the Great _____.

9. The _____ River is the longest river in the United States.

10. The Great _____ contain just over 20 percent of the world's fresh water.

Activities

Make some food from a traditional Thanksgiving feast with your parent/teacher. This might include turkey, cranberry sauce, pumpkin pie, rolls, and more!

Short Answers

1. What does the green on the Mexican flag signify?

2. What is the capital city of Mexico?

3. In Mexico they use the Mexican _____ as a monetary unit.

4. The priest Diego Carranza served the _____ Indians for 12 years to tell them of Jesus.

5. As many as a billion butterflies travel from as far as Canada to go to the Monarch Butterfly _____ Reserve.

6. _____ monsters are poisonous lizards covered in colored scales.

7. _____ can be hot or cold, but most often are dry, receiving less than 10 inches of rain each year.

8. The _____ cactus can grow over 60 feet tall.

9. Spanish soldiers led by Fernando Cortez were the first _____ to see Tenochtitlan.

10. Chichen _____ was a thriving city over a thousand years ago.

Activities

Make some guacamole dip with your parent/teacher.

Ingredients

3 ripe avocados (ripe as in really dark in color and slightly soft when squeezed)

1 small onion, finely diced

1 medium tomato, diced

1 lime (juice of)

salt and pepper (to taste)

Tabasco sauce (optional)

Directions

1. Take avocados and cut into and around the circumference of the avocado's pit.
2. With both hands twist the two halves in opposite directions and pull apart.
3. Remove the pit with a knife or a spoon.
4. Carve out the flesh of the avocado with a tablespoon.
5. Place avocados into a medium bowl and mash and stir until moderately creamy. I use a fork or a pastry blender.
6. Some people like the avocados more chunky than others — either way it tastes great.
7. Add diced onion and tomatoes to the avocados.
8. Add salt, pepper, Tabasco sauce, and lime juice and taste to see if seasoned enough to your liking.
9. In a separate small bowl, put in 2 tablespoons of the dip.
10. Add 1 tbsp of sour cream and adjust your seasonings again.
11. Taste it and decide if you prefer your guacamole with or without sour cream.
12. If you like with sour cream, add 4 tablespoons or more to the avocado mixture and adjust your seasonings again.
13. If not, do nothing and leave it like it is.
14. Serve chilled with chips or add a little to your taco salad, tacos, or fajitas.

From food.com

Short Answers

1. What are the primary languages of Canada?

2. What is the capital city of Canada?

3. Canada is made up of _____ provinces and _____ territories.

4. The sport of lacrosse was first named this by a _____ missionary.

5. The Canadian Bill of Rights was founded "upon principles that acknowledge the supremacy of _____."

6. It is believed that Niagara Falls was formed by melting Ice Age _____.

7. It is likely the Inuit people here crossed a bridge of _____ formed after the Great Flood of Noah.

8. God designed the coat of the _____ fox to change to snow white in the winter.

9. The five major agricultural industries of Canada are grains, beef and hogs, dairy, poultry and eggs, and _____.

10. In Nova Scotia one can see the world-famous fossil cliffs on the Bay of _____.

Activities

Make some Canadian poutine with your parent/teacher.

Ingredients

1 quart vegetable oil (for frying)

1 (10¼ ounce) can beef gravy

5 medium potatoes, cut into fries

2 cups cheese curds

Directions

1. Heat oil in a deep fryer or deep heavy skillet to 365°F (185°C).
2. Warm gravy in saucepan or microwave.
3. Place the fries into the hot oil, and cook until light brown, about 5 minutes.
4. Remove to a paper towel-lined plate to drain.
5. Place the fries on a serving platter, and sprinkle the cheese over them.
6. Ladle gravy over the fries and cheese, and serve immediately.

From food.com

Short Answers

1. The _____ is the longest mountain range in the world.

2. Dutch explorer Jacob Roggeveen named the island of Rapa Nui _____ Island in honor of the sacred day of Christ's Resurrection.

3. _____ is the tallest mountain on the continent of South America.

4. The highest uninterrupted waterfall in the world is Angel Falls in _____.

5. What is the capital city of Brazil?

6. What is the primary language of Brazil?

7. The _____ River flows to the Atlantic Ocean and is the second-longest river in the world.

8. _____ are the world's largest rodents.

9. Christianity was introduced to Brazil as far back as the _____.

10. _____ are the largest snakes in the world.

Activities

Make some churrasco with your parent/teacher.

Ingredients

3 heads garlic (about 30 to 40 cloves)

2 teaspoons salt

1 teaspoon black peppercorns

1 cup orange juice

¼ cup fresh lime juice

¼ cup fresh lemon juice

1 cup onion, minced

2 teaspoons oregano

1 cup Spanish olive oil

meat, of choice pounded or cut thin (beef, chicken, pork)

Directions

1. Mash garlic, salt, and peppercorns into a paste, using a mortar and pestle. Stir in orange juice, lime juice, lemon juice, onion, and oregano. Let sit at room temperature for 30 minutes or longer. Whisk the garlic-orange juice mixture with the olive oil until well blended.

2. Preparing the meat: Preparing the meat is very easy. Place the meat in a large bowl or pan. Add enough marinade to cover the meat. Place a cover over the bowl or pan and place in the refrigerator a minimum of five hours, but preferably overnight.

3. Grill the marinated meat outdoors on the barbecue. (Make sure your coals are hot and white!) For the steak — you can cook it however you like it — from rare to well done. However, the rarer the meat, the more tender and flavorful!

From food.com

Short Answers

1. What does the red on the Bolivian flag stand for?

2. What is the capital city of Bolivia?

3. The monetary unit used in Bolivia is the _____.

4. Maned wolves are sometimes called _____ wolves because of their bad odor.

5. The _____ people live upon constructed, floating islands made of bundled reeds.

6. Pumapunku is a mysterious site featuring block-like, large _____ with precise holes and edges.

7. The Christian message was first introduced as far back as _____.

8. Christmas Eve in Bolivia is called Misa del Gallo, or mass of the _____.

9. Homes made of brick or cement are safer than _____ homes because of disease spread by the vinchuca bug.

10. Lake Titicaca has _____ main rivers flowing into it.

Activities

Make some Bolivian salsa cruda with your parent/teacher.

Ingredients

1 cup onion, finely chopped (preferably sweet onion)

1 cup tomato, ripe, peeled, finely chopped

1 hot pepper, seeds removed, finely chopped

1 tablespoon parsley, very finely chopped

1 teaspoon salt (to taste)

1 teaspoon black pepper, coarse-ground

½ tablespoon oil (virgin olive oil preferred)

Directions

1. Mix all the ingredients, drizzling over the oil last.
2. If you like, add a squeeze of lemon juice and some finely chopped garlic.
3. The sauce improves in flavor if covered and left to stand for a while.

From food.com

Short Answers

1. Mount _____ is the highest mountain on the continent of Europe.

2. The smallest island in the world is _____ Rock.

3. _____ City is the smallest independent country in the world.

4. What is the capital city of Norway?

5. What is the monetary unit used in Norway?

6. The northern lights are often called aurora from the Latin word that means "_____."

7. The traditional Norwegian clothing of the 18th to 20th centuries is called _____.

8. Along the west coast of Norway are thousands of _____ or inlets, waterways carved by glaciers.

9. On December 13, Christians in Norway celebrate the festival of St. _____.

10. Olav Tryggvason became the first Christian _____ of Norway over a thousand years ago.

Activities

Make some lussekatter with your parent/teacher.

Ingredients (metric)

300 ml milk

1 g saffron

50 g baker's yeast

150 g sugar

125 g butter or 125 g margarine

700 g all-purpose flour

1 egg

salt

raisins

Directions

1. Melt butter or margarine in a pan and add the milk and the saffron.
2. Warm the mixture to 37°C (100°F).
3. Use a thermometer; the correct temperature is important!
4. Pour the mixture over the finely divided yeast; then add the remaining ingredients (except for the egg and the raisins), which should have a temperature of 21–23°C (72–75°F).
5. Mix into a smooth dough.
6. Cover the dough with a piece of cloth and let it rise for 30 minutes.
7. Knead the dough, divide it into 25–30 pieces and form each piece into a round bun.
8. Let the buns rest for a few minutes, covered by a piece of cloth.
9. Form each bun into a string, then arrange the string in a suitable shape, e.g., an S or double S. Regardless of the shape, the ends of the string should meet.
10. Press a few raisins into the dough.
11. Cover the "Lucia cats" with a piece of cloth and let them rise for 40 minutes.
12. Whip the egg with a few grains of salt, and paint the "Lucia cats" with the mixture.
13. Bake them for 5–10 minutes in the oven at 250°C (475°F) until golden brownish yellow.

From food.com

Short Answers

1. What is the capital city of Italy?

2. Renaissance is a word that means "_____."

3. What kind of government system does Italy have?

4. The Italian flag was inspired by the _____ flag brought to Italy in 1797.

5. What kind of monetary unit is used in Italy?

6. Leonardo da Vinci left a legacy with his paintings such as *The Last* _____ and the *Mona Lisa*.

7. Swordfish use their "swords" to _____ their prey to be able to catch them.

8. St. Peter's Basilica was originally built in the fourth century to hold the _____ tomb.

9. _____ the Great made Christianity the official religion of the Roman Empire.

10. The Roman Colosseum was completed in A.D. _____.

Activities

Make some Italian breadsticks with your parent/teacher.

Ingredients

1 cup warm water

3 tablespoons soft butter

1½ teaspoons salt

3 cups bread flour

2 tablespoons sugar

1 teaspoon Italian seasoning

1 teaspoon garlic powder

2¼ teaspoons active dry yeast

Topping

1 tablespoon melted butter

1 tablespoon grated parmesan cheese

Directions

1. Place all dough ingredients in your bread machine pan according to manufacturer's direction.
2. Select dough setting.
3. When dough cycle is complete, turn out onto floured board and divide dough in half.
4. Cut each portion into 12 pieces and roll each piece into a 4 to 6" rope.
5. Place 2" apart onto greased baking sheets.
6. Cover and let rise in a warm place for about 20 minutes or until doubled.
7. Bake at 350° for 15 minutes or until golden brown.
8. Immediately brush with butter and sprinkle with Parmesan cheese.
9. Note: Be careful not to overbake or they will become tough.

From food.com

Short Answers

1. What is the capital city of France?

2. In 1940, the caves of _____ were discovered, covered in colorful paintings.

3. What is the national symbol of France?

4. _____ was influential in bringing her husband, King Clovis I, to Christianity.

5. The cathedral of _____ is a magnificent work of art from the 12th century.

6. The _____ de France is a bicycle race first held in 1903.

7. From 350 to 400 different kinds of _____ are produced in France.

8. The painter Vincent van Gogh had a deep love for the _____ people.

9. The plant _____ is used to develop honey flavor, as well as a scent for soaps.

10. The _____ Tower is one of the most recognizable structures from the modern age.

Activities

Make some Gateaux with your parent/teacher.

Ingredients (metric)

500 g butter

225 g cornstarch

3 egg yolks

500 g flour

1 teaspoon baking powder

250 g sugar

Directions

1. In a bowl, beat butter, egg yolks, and sugar until light and well blended.
2. In a separate bowl, sift together cornstarch, flour, and baking powder. Gradually add the dry mixture to the wet mixture, blending well between each addition.
3. Use a pastry tube with a medium-sized start mold to make small circle (wreath type) cookies.
4. Bake in medium-hot or hot oven until cookies are lightly browned.

From food.com

Short Answers

1. What does the design of the German flag relate to?

2. What is the capital city of Germany?

3. Cars can travel up to 200 miles per hour on Germany's _____.

4. In 1456, Johannes Gutenberg printed the _____ in Mainz.

5. One of the Protestant Reformation's refrains was *Sola Gratia*, which means _____.

6. Johann Sebastian Bach often wrote SDG at the end of his music, which meant _____.

7. One of the most famous ballets in the world is the _____, adapted from the German by Russian composer Tchaikovsky.

8. Neuschwanstein Castle built by Ludwig II of Bavaria, means "New _____ _____."

9. The world's fresh water is kept pure through the water _____.

10. Many Christmas tree traditions are linked to Martin _____ in Germany.

Activities

Make some stollen with your parent/teacher.

Ingredients

1 (¼ ounce) package active dry yeast

¾ cup warm water (105–115°)

½ cup sugar

½ teaspoon salt

3 eggs

1 egg, separated

½ cup butter, softened

3½ cups flour

½ cup blanched almonds, chopped

¼ cup citron, cut up

¼ cup candied cherry, cut up (optional)

¼ cup raisins

1 tablespoon lemon peel, grated

1 tablespoon butter, softened

1 tablespoon water

Glaze

1½ cups powdered sugar

4½ teaspoons milk

Directions

1. Dissolve yeast in ¾ c water in a large bowl. Beat in sugar, salt, eggs, egg yolk, ½ c butter, and 1¾ c flour with an electric mixer on medium speed 10 minutes, scraping bowl constantly. Stir in remaining flour, almonds, citron, cherries, raisins, and lemon peel. Scrape batter from side of bowl. Cover and let rise in warm place 1½ to 2 hours or until doubled. Dough is ready if indentation remains when touched. Cover and refrigerate egg white.

2. Stir down batter by beating about 25 strokes. Cover tightly and refrigerate at least 8 hours or overnight.

3. Grease cookie sheet. Turn dough onto well-floured surface; turn to coat with flour. Divide in half. Press each half into an oval 10 x 7". Spread with butter. Fold ovals lengthwise in half; press only folded edge firmly. Place on cookie sheet. Beat egg white and 1 tablespoon water; brush over folded ovals. Cover and let rise 45–60 minutes or until doubled.

4. Heat oven to 375°. Bake 20–25 minutes or until golden brown. While warm, dust with powdered sugar or spread with glaze.

5. To make glaze, mix powdered sugar and milk until smooth and spreadable.

From food.com

Short Answers

1. What is the capital city of the United Kingdom?

2. What type of monetary unit do they use in the United Kingdom?

3. Name the national symbols for England _____, Scotland _____, Wales _____, and Northern Ireland _____.

4. The Old Course at _____ is considered by many to be the oldest golf course in the world.

5. By his direct efforts, William _____ helped get the slave trade outlawed in the United Kingdom in 1807.

6. King James I of England commissioned an English Bible translation in 1604 that was completed in _____.

7. Thousands of _____ flock to Scotland's coastlines.

8. The _____ _Tales_ were written by Geoffrey Chaucer and first printed in 1475.

9. _____ National Park is the first and largest national park in Wales.

10. The _____ Causeway is an area in Northern Ireland with 40,000 interlocking columns of basalt.

Activities

Make some Bara Brith with your parent/teacher.

Ingredients

6 ounces dried fruit

8 ounces dark brown sugar

½ pint strong hot tea

10 ounces self-riising flour

1 egg

Directions

1. Soak the dried fruit and sugar overnight in the tea.
2. You can use either fresh tea or the cold dregs from the teapot (this gives a good strong color). Next day, sieve the flour and fold it into the fruit. Mix in the lightly beaten egg.
3. Line a small loaf-tin with buttered paper then tip in the mixture, smoothing it well into the corners.
4. Bake in a gentle oven at 300°F (150°C) for 1½ hours.
5. Cool and store for at least 2 days in a tin so that it matures moist and rich.
6. Traditionalists say you should never butter the Bara Brith, but it's lovely that way!

From food.com

Short Answers

1. What is the capital city of Russia?

2. What is the monetary unit used in Russia?

3. The longest rail system in the world is the _____-_____ Railway.

4. Contrary to rumors, Yuri Gagarin never stated that he didn't see _____ when he was in space.

5. The Russian _____ Bible was the first full Bible translation of Scripture into Russian.

6. Tsar _____ the Great laid the foundations for the city of St. Petersburg.

7. Russia covers a total of _____ time zones.

8. The _____ seal is found only in the waters of Lake Baikal in Siberia.

9. The arctic _____ is a permanently frozen landscape.

10. The Russian language is written in characters called the _____ alphabet, named after St. Cyril.

Activities

Make some bliny with your parent/teacher.

Ingredients

½ cup water, lukewarm

½ cup buckwheat flour

2 cups milk, warmed slightly

½ teaspoon salt

½ lb butter, melted & cooled

3 egg whites

1.5 (⅓ ounce) packages yeast

2 cups flour, white

3 egg yolks, slightly beaten

1 teaspoon sugar

3 tablespoons sour cream

16 ounces black caviar or 16 ounces lox

Directions

1. Dissolve yeast in water and let sit for 3–5 minutes.
2. In a large bowl, mix ¼ cup buckwheat flour and 2 cups white flour.
3. Make a well in the flour and pour in the yeast mixture.
4. Slowly mix with a wooden spoon, then beat mixture until smooth.
5. Cover and let sit in warm spot for 3 hours.
6. Stir vigorously and add the rest of the buckwheat flour.
7. Cover and let sit in warm spot for another 2 hours.
8. Stir and slowly add milk, egg yolks, salt, sugar, 3 tbsp butter, and 3 tbsp sour cream.
9. In a separate bowl, beat egg whites until stiff.
10. Fold egg whites into dough.
11. Cover and let sit for another 30 minutes.
12. Coat the bottom of a frying pan with butter and heat.
13. Place 3 tbsp batter in pan and fry 2–3 minutes.
14. Brush with butter and flip, and fry other side 2–3 minutes.

From food.com

Short Answers

1. _____ Falls is on the Zambezi River, and has been called "the smoke that thunders."

2. _____ is the highest mountain on the continent of Africa.

3. The largest artificial lake or reservoir in the world is Lake _____.

4. 3The world's longest river is the _____.

5. David _____ was a missionary and the first European to explore the interior of Africa.

6. What is the capital city of Kenya?

7. What monetary unit is used in Kenya?

8. The largest lake in Africa is Lake _____.

9. Giant _____ eat termites and ants.

10. In Africa, grasslands are specifically called _____.

Activities

Make some fufu (ugali) with your parent/teacher.

Ingredients

2½ cups Bisquick

2½ cups instant potato flakes

Directions

1. Bring 6 cups of water to a rapid boil in a large, heavy pot.

2. Combine the two ingredients and add to the water.

3. Stir constantly for 10–15 minutes — a process that needs two people for best results: one to hold the pot while the other stirs vigorously with a strong implement (such as a thick wooden spoon). The mixture will become very thick and difficult to stir, but unless you are both vigilant and energetic, you'll get a lumpy mess.

4. When the fufu is ready (or you've stirred to the limits of your endurance), dump about a cup of the mixture into a wet bowl and shake until it forms itself into a smooth ball.

5. Serve on a large platter alongside a soup or stew.

From food.com

Short Answers

1. What is the capital city of Egypt?

2. What is the primary language of Egypt?

3. The national symbol of Egypt is the _____ eagle.

4. The ancient name of the country in Egyptian is _____, which means "black land."

5. The entire country of Egypt is located within the _____ Desert.

6. Many of the desert-dwelling tribes who live in tents or other moveable structures are called _____.

7. The Pyramids at _____ are the only surviving relic of the Seven Wonders of the Ancient World.

8. Samuel Zwemer was a missionary who felt a calling to reach _____ with the gospel.

9. Joseph, Mary, and their young child Jesus fled to Egypt to escape _____ wrath.

10. The majority of Christians in Egypt are called _____ Christians.

Activities

Make some Egyptian red lentil soup with your parent/teacher.

Ingredients

5 cups vegetable broth or 5 cups water

1 cup dried red lentils

2 cups chopped onions

2 cups chopped potatoes

8 garlic cloves, peeled and left whole

1 tablespoon canola oil

2 teaspoons ground cumin

½ teaspoon turmeric

1 teaspoon salt

⅓ cup chopped fresh cilantro

3 tablespoons fresh lemon juice

salt and pepper

Directions

1. Add the first 5 ingredients to a large pot; cover and bring to a boil.

2. Lower the heat and simmer 15–20 minutes or until the lentils and veggies are tender.

3. Take pot from stove burner and set aside.

4. In a small saucepan, add the oil; warm over low heat until the oil is hot but not smoking.

5. Add in the cumin, turmeric, and salt; cook and stir constantly for 2–3 minutes or until the cumin has released its fragrance (be careful not to scorch the spices).

6. Set spice mixture aside for 1 minute to cool.

7. Stir spice mixture into the lentil mixture; add cilantro, stir to combine.

8. You can puree the soup, in batches, in a blender OR you can use an immersion blender and blend to desired texture (I like to leave it a little chunky).

9. Add in lemon juice; stir to combine.

10. Rewarm soup in soup pot; season if needed with salt/pepper.

From food.com

Short Answers

1. The Y in the South African flag stands for "taking the road ahead in _____."

2. What is the administrative capital city of South Africa?

3. What kind of governmental system exists in South Africa?

4. The 19th-century discovery of _____ and gold brought many more settlers to the area.

5. Antonio de Saldanha carved a _____ into the Lion's Head rock in 1503.

6. _____ baboons are considered to be endangered in some areas of South Africa.

7. Robert Moffat translated both the Bible and *The Pilgrim's Progress* into _____.

8. The _____ Mission Station was a missions and educational institution founded in 1824.

9. Cape of Good _____ was a city where people came to escape religious persecution.

10. The southern tip of the continent is Cape _____.

Activities

Make some samosas (samoosas) with your parent/teacher.

Ingredients

1 lb lamb or 1 lb hamburger

5 potatoes

2 tablespoons cilantro

½ teaspoon onion, chopped

½ teaspoon curry powder

½ teaspoon cinnamon

salt

Directions

1. Wash meat if using lamb, dry and cook.
2. Break into small pieces.
3. Boil potatoes and cut into small pieces.
4. Mix meat, potato, cilantro, onion, and seasonings.
5. Place small amount in corner of an egg roll wrapper; fold wrapper in.
6. Half so it is triangular in shape.
7. Use milk to seal.
8. Deep fry.

From food.com

Short Answers

1. Asia is the largest continent at nearly _____ million square miles.

2. Mount _____ in Nepal is the highest mountain on earth.

3. Built perhaps as early as 300 years before the birth of Christ, _____ is a remarkable stone city.

4. Istanbul, Turkey, is a city on the two continents of _____ and Asia.

5. In the Hindi language, the word for bear is _____, and the inspiration for the name of the bear in *The Jungle Book*.

6. What is the capital city of Israel?

7. The _____ Sea or Salt Sea is the earth's lowest elevation on land.

8. Jerusalem has been given the name City of _____.

9. The _____ is Israel's national bird.

10. _____ is a Hebrew word that means "community settlement."

Activities

Make some Jewish burekas with your parent/teacher.

Ingredients

Dough

3 cups flour

1 cup margarine

½ cup water

½ cup vegetable oil

⅛–¼ teaspoon salt

Filling

4 or 5 medium white potatoes, cooked, peeled, and mashed

¼ lb asiago cheese, grated (about 1 cup)

1½ cups farmer cheese

4 large eggs

Egg Wash

1 large egg, beaten with 1 tablespoon water

Directions

1. Beat the flour and margarine together, using a hand-held or upright mixer.
2. Add the water, oil, and salt and continue beating until well combined.
3. By hand, blend in any flour still remaining at the bottom of the mixing bowl.
4. Refrigerate the dough for at least 1 hour.
5. Combine the filling ingredients, using a hand-held or upright mixer.
6. Take the dough out of the refrigerator.
7. Separate into 24 equal portions.
8. Pat each portion of the dough into a circle approximately 4 inches in diameter and dust with flour.
9. Place a scant tablespoon of filling on one side of each circle.
10. Fold the unfilled portion over the filling.
11. Transfer borekas onto a non-stick baking sheet.
12. Crimp the edges with a fork.
13. Using a pastry brush, "paint" the top of each boreka with egg wash.
14. Bake in a preheated 375° oven for 35 minutes.

From food.com

Short Answers

1. What is the capital city of Saudi Arabia?

2. What type of government system do they have?

3. What kind of monetary unit do they use?

4. The city of _____ is considered the holiest city of Islam.

5. Millions come to Mecca in what is called the _____, a pilgrimage.

6. Saudi Arabia has _____ major deserts.

7. Arabic _____ is the beautiful writing style of the region that originated in the 6th century.

8. The _____ Church was one of the first church buildings ever constructed.

9. No religion other than _____ is allowed to be practiced in Saudi Arabia.

10. The numbers of elegant Arabian _____ are once again growing.

Activities

Make some kebabs (kababs) with your parent/teacher. Select beef, chicken, or shrimp, and add your favorite veggies as well. These can be done in the oven or outside on a grill!

Short Answers

1. The flag of India has a 24-spoked wheel called a _____.

2. What is the capital city of India?

3. What monetary unit is used in India?

4. The longest river in India is the _____.

5. Shah Jahan had the Taj _____ built as a monument for his wife who had passed away.

6. Mother _____ came to serve the poor in India and started the Missionaries of Charity.

7. William _____ helped complete translations of the Bible into Bengali, Marathi, and Sanskrit.

8. _____ is the most widely spoken language and primary tongue of 41 percent of the people.

9. The Hindu festival of Raksha Bandhan is celebrated in August, and the name means "_____ of protection."

10. One of the rarest predators in the world is the _____ leopard.

Activities

Make some chana with your parent/teacher.

Ingredients

1 (15 ounce) can chickpeas

1 onion, chopped

1 tomato, chopped

1 green chili pepper, chopped

4–5 garlic cloves, chopped

1 inch ginger root, chopped

2–3 bay leaves

1 teaspoon red chili powder

½ teaspoon turmeric powder

1 teaspoon coriander powder

1 teaspoon garam masala powder

3 tablespoons olive oil

coriander leaves, for garnishing

Directions

1. Cut onion, tomato, and green chili. Grind it in food processor along with ginger and garlic and make paste.
2. Heat oil in a pan and fry bay leaves for 30 secs.
3. Add the paste and fry on medium heat until golden brown (the oil starts separating from the mixture).
4. Add red chili powder, turmeric powder, coriander powder, garam masala, and salt. Mix well. Fry for 2–3 minutes.
5. Add enough water to make thick gravy. Bring the gravy to a boil.
6. Add the can of chick peas. Stir well and cook over medium heat for 5–7 minutes.
7. Garnish with chopped green coriander leaves and serve hot.

From food.com

Short Answers

1. What is the capital city of China?

2. What is the national symbol of China?

3. The longest man-made waterway in the world is the _____ Canal.

4. The tomb of China's first emperor was filled with almost 8,000 _____ figures.

5. Christians entered the capital of the _____ Dynasty in A.D. 635.

6. Hudson _____ adopted many Chinese customs to connect the gospel to the people.

7. The Three Gorges Dam on the _____ River was completed in 2012.

8. Written Chinese does not include an _____.

9. _____ has been produced in China for over 4,000 years.

10. The most important holiday for the people of China is the Chinese _____ Year.

Activities

Make some rice with your parent/teacher. You can steam simple white rice or make up some fried rice, whichever your family likes best.

Short Answers

1. The Japanese characters that make up the name of the country mean "sun _____."

2. What is the capital city of Japan?

3. Japan is an _____ or chain of islands.

4. _____ means "floating world pictures" and refers to the temporary nature of beauty in the world.

5. When Christianity was outlawed, the beliefs of Christ were kept by the _____ Kirishitan or "hidden Christians."

6. _____ is a word that literally means "tray plantings."

7. The art form of _____ means "folding paper."

8. Mount _____ is an active volcano on Honshu Island.

9. _____ Day or Kodomo no Hi is celebrated on May 5.

10. The Japanese _____ or goat-antelope is found in the thick forests of Honshu.

Activities

Make some yakitori with your parent/teacher.

Ingredients

1½ cups mirin

¾ cup soy sauce

4 tablespoons sugar

1 garlic clove, pressed (optional)

1 lb boneless chicken thighs

1–2 green onions, in inch-long pieces (optional)

8 bamboo skewers

Directions

1. Put mirin, soy sauce, sugar, and garlic, if using, into a medium-sized sauce pan, and cook over medium heat until there is only half as much sauce remaining.
2. Let cool slightly while you are preparing the chicken.
3. Cut chicken into bite-sized pieces; keeping the skin on is usual in Japan.
4. Thread chicken onto skewers, alternating with green onion if desired.
5. Begin grilling the chicken skewers on both sides without the sauce, when the meat starts changing color, brush the sauce on both sides, and continue grilling, brushing on sauce about 3 times total, and turning until done.

From food.com

Short Answers

1. What is the capital city of Malaysia?

2. What kind of governmental system does Malaysia have?

3. What is the national symbol of the nation?

4. Malaysia is made up of _____ states and three territories.

5. The _____ Towers are some of the tallest buildings in the world.

6. The largest identified cave chamber in the world is the _____ Chamber in the Gunung Mulu National Park.

7. The _____ is one of the largest flowers in the world.

8. _____ traders in the 7th century included some of the first Christians to the area.

9. The _____ Mission Society began their ministry to the migrant Chinese residents in 1882.

10. The _____ bear or honey bear is nocturnal, eating at night and sleeping in the day.

Activities

Make some rempah with your parent/teacher.

Ingredients

⅛ cup cashews

1 onion

3 garlic cloves

1 teaspoon chili powder

1 teaspoon ground turmeric

1 teaspoon ground coriander

1 tablespoon miso (original recipe calls for anchovy paste — use that if you like to make it authentic)

1 tablespoon chopped fresh lemongrass

1 inch fresh ginger, chopped

1 tablespoon vegetable oil

Directions

1. Add all ingredients to a small food processor. Process until almost smooth.

2. There'll be some small pieces but don't worry, you'll cook everything soft when you use it to make a curry or whatever you're making. You just need a little oil and a hot pan.

3. You can add in some coconut milk to make it richer, thin out with water if you need to. Add veggies and, if you wish, meat or fish and you have a flavorful dish!

From food.com

Short Answers

1. The _____ devil is a fierce hunter and scavenger, about the size of a small dog.

2. Mount _____ is the highest mountain on the continent of Australia.

3. _____ are formed when various algae, clams, corals, sponges, and other creatures cement themselves together in warm, shallow seas.

4. The world's largest living reptile is the _____ crocodile.

5. What is the capital city of Australia?

6. The design of the _____ Opera House was inspired by great sailing ships.

7. The River _____ is the longest river in Australia.

8. From a prayer book his wife gave him, Captain James _____ named many of the places on the coast of Australia.

9. The _____ bats or flying foxes don't use echolocation to find their food.

10. The aboriginal people still make an ancient instrument called a _____.

Activities

Make some Australian pikelets (pancakes) with your parent/teacher.

Ingredients

1 large egg

¼ cup sugar (use vanilla sugar if you have it, or as some reviewers have suggested, add ½ to 1 teaspoon of vanilla extract)

½ cup milk (1–2 tbsp extra may be required for correct consistency)

1 cup self-rising flour

Directions

1. In a pouring jug, whisk together well the egg, sugar, and half of the milk.
2. Add the flour and mix thoroughly.
3. Add milk until the batter is the consistency of a thick cream.
4. Beat until smooth.
5. Pour about a soup spoon sized amount onto a preheated lightly greased frypan (you will need to start about medium-high heat but after a few rounds you will need to drop it back to medium).
6. Flip with a spatula when the top is all bubbly (should be a lovely golden color), and cook the other side.
7. Remove from pan and pile piklets up on a plate (cover with a paper towel so they don't go rubbery).
8. Serve with jam, honey, or maple syrup, etc.

From food.com

Short Answers

1. The stars on the flag of New Zealand represent the Southern _____ constellation.

2. What is the capital city of New Zealand?

3. _____ are narrow bodies of water carved by glaciers, which then filled with water.

4. About the size of a chicken, the flightless _____ bird is a national symbol of New Zealand.

5. _____ are the indigenous or native people of New Zealand.

6. _____ originally came from China, where New Zealander Mary Isabel Fraser had been visiting a mission school.

7. Rudyard Kipling called _____ Sound the eighth wonder of the world.

8. Mount _____ is the tallest mountain in New Zealand.

9. _____ is a geothermal region on the North Island.

10. What is the primary language of New Zealand?

Activities

Eat some kiwi fruit with your parent/teacher.

Short Answers

1. What is the name of the largest of Antarctica's subglacial lakes?

2. _____ penguins are the largest of the flightless birds here.

3. Up to _____ percent of the earth's ice is found in the Antarctic ice cap.

4. The _____ midge is the largest land animal here.

5. Mount _____ is the highest mountain on the continent of Antarctica.

6. It's so cold in various places of the Antarctic that it is known as the world's _____ desert.

7. The Antarctic _____ are tiny creatures that thrive in the cold waters.

8. The _____ Lutheran Church on Grytviken first opened on Christmas Day 1913.

9. The first team to reach the South _____ was led by Roald Amundsen.

10. _____ Church holds about 30 worshipers and opened on King George Island in 2004.

Children's Atlas of God's World

Quiz

Q	*Children's Atlas of God's World* Concepts & Comprehension	Quiz	Scope: Biomes of the World	Total score: ____ of 100	Name

Answer Questions: (10 Points Each Question)

1. Biomes is a shortened form of _____ homes.

2. The dry side of a mountain is called the "rain _____."

3. Brooks, creeks, and streams are fed by rainfall and _____.

4. _____ are found where there is too little rain for trees and more rain than is found in a desert.

5. _____ are places that receive less than 10 inches of rain per year.

6. The soil quality in rainforests is actually very _____.

7. Deciduous forests lose their _____ in fall.

8. _____ reefs are the biggest reefs in the ocean.

9. _____ are formed either by volcanoes or the movement of the earth's crust.

10. Warmer months on the _____ are actually filled with wondrous life.

Answer Keys

Answer Keys

Passport to the World ⚷ Worksheet Answer Keys

Armenia
1. Yerevan
2. *Parev*
3. *Tsedesutyun*
4. *Shnorhagallem*
5. *Khanaghutyun*
6. Six million
7. two heads
8. Ararat
9. chasharan
10. Turkish

Bangladesh
1. Dhaka
2. *Nomoskar*
3. *Accha*
4. *Dhanyabad*
5. *Shanti*
6. 200 million
7. bread
8. rhymed
9. mangrove
10. 80

United States of America
1. Washington D.C.
2. *Ohseeyo*
3. *Donadagovi*
4. *Wado*
5. *Dohiyi*
6. 20,000
7. shared
8. 8.5
9. Oklahoma or North Carolina
10. 40

Netherlands
1. Amsterdam
2. *Daag* or *geode dag*
3. *Tot ziens*
4. *Dank u*
5. *Vrede*
6. 17 million
7. friend
8. fourth
9. shoes
10. herring

Australia
1. Canberra
2. Hello
3. Goodbye
4. Thank you
5. Peace
6. 340 million
7. book
8. 380,000
9. half
10. didgeridoo or didge

France
1. Paris
2. *Bonjour*
3. *Au revoir*
4. *Merci*
5. *Paix*
6. 50 million
7. spice
8. Eiffel
9. 300
10. England

Canada
1. Ottawa
2. *Bonjour*
3. *Au revoir*
4. *Merci*
5. *Paix*
6. 50 million
7. Patience
8. lacrosse
9. Toronto
10. three

Germany
1. Berlin
2. *Guten tag*
3. *Auf wiedersehen*
4. *Danke*
5. *Frieden*
6. 75 million
7. honest
8. Ulm
9. Mainz
10. pretzels

Israel
1. Jerusalem
2. *Shalom*
3. *Lehitrahott*
4. *Toda*
5. *Shalom*
6. 5 million
7. little
8. Dead or Salt
9. Old
10. braille

Iceland

1. Reykjavik
2. *Góðan dag*
3. *Bless*
4. *Takk*
5. *friður*
6. 308,900
7. bookless
8. 130
9. horse
10. poppies

Japan

1. Tokyo
2. *Konnichiwa*
3. *Sayonara*
4. *Arigatou*
5. *Heiwa*
6. 122,000,000
7. virtue
8. pyramid
9. earthquakes
10. Rice

South Korea

1. Seoul
2. *Annyeong*
3. *Annyong-hi kashipshio*
4. *Kamsahamnida*
5. *Phyongh'wa*
6. 48,000,000
7. dirt
8. October
9. A South Korean won
10. Kimchi

Lithuania

1. Vilnius
2. *Labas*
3. *Viso gero*
4. *Achiu*
5. *Taika*
6. 3 million
7. love
8. litai
9. salt
10. 12 (one for each of the 12 Apostles)

China

1. Beijing
2. *Ni hăo*
3. *Zàijìan*
4. *Xìexìe*
5. *He ping*
6. 1,330,000,000
7. gold
8. Backbone
9. 200
10. renminbi yuan

Norway

1. Oslo
2. *Goddag*
3. *Ha det bra*
4. *Takk*
5. *Fred*
6. 4,676, 305
7. grass
8. Norwegian kroner
9. 1,600
10. Henrik

India

1. New Delhi
2. *Namaskar*
3. *Vidaaya*
4. *Dhanyabahd*
5. *Shanty*
6. 31 million
7. crocodile
8. Indus
9. universities
10. Indian rupee

Afghanistan

1. Kabul
2. *Salam aleikum*
3. *Da khoday-pe-aman*
4. *Tashakor*
5. *Amniat*
6. One million
7. strength
8. Kite
9. Afghanis
10. Dari

Bolivia

1. La Paz
2. *Raphi*
3. *Tupananchiskama*
4. *Yusulpayki*
5. *Qasikay*
6. Over two million
7. lazy
8. temple
9. Incas
10. yerba

Russia

1. Moscow
2. *Zdravstvujte*
3. *Dos vidanija*
4. *Spasibo*
5. *Mir*
6. Russian rubles
7. cry
8. largest
9. honey
10. week

Mexico

1. Mexican pesos
2. *Hola*
3. *Adios*
4. *Gracias*
5. *Paz*
6. 350,000,000
7. mind
8. Aztec
9. surname
10. Mexico City

Turkey

1. Ankara
2. *Merhaba*
3. *Gule gule*
4. *Tesekkur ederim*
5. *Sulh*
6. Turkish liras
7. good
8. rock
9. Istanbul
10. kahve

Ukraine

1. Kyiv
2. *Vitayu*
3. *Do pobachennya*
4. *Dyakuju*
5. *Mir*
6. Hryvnia
7. lemon
8. Soviet Union
9. bread
10. pysanka

Vietnam

1. Hanoi
2. *Chao anh* (to a man) or *chao chi* (to a woman)
3. *Tam biet*
4. *Cam on*
5. *Hoa binh*
6. Dong
7. learning
8. 938
9. Food
10. missionaries

United Kingdom

1. London
2. *Shwmae*
3. *Hwyl fawr*
4. *Diolch*
5. *Heddwich*
6. 575,000
7. language
8. Jurassic
9. tea
10. Windsor

South Africa

1. Cape Town
2. *Molo*
3. *Sala kakuhle*
4. *Enkosi*
5. *Uxolo*
6. 7,000,000
7. swallowing
8. gold
9. rand
10. three

Nigeria

1. Abuja
2. *E ku aaro*
3. *O dabo*
4. *E se*
5. *Alaafia*
6. 19,000,000
7. horse
8. Nairas
9. Zuma
10. ayo

Swaziland

1. Mbabane
2. *Sawubona*
3. *Hamba hahle*
4. *Ngiyabona*
5. *Ukuthula*
6. Emalangeni
7. customs
8. beehives
9. Mswati
10. Norway

United States

1. Challenger
2. McKinley
3. Greenland
4. Washington, D.C.
5. *The Star-Spangled Banner*
6. Creator
7. needles
8. Flood
9. Missouri
10. Lakes

Mexico

1. Hope, joy, and love
2. Mexico City
3. peso
4. Chontal
5. Biosphere
6. Gila
7. Deserts
8. saguaro
9. Europeans
10. Itza

Canada

1. English and French
2. Ottawa
3. ten, three
4. Jesuit
5. God
6. glaciers
7. ice
8. Arctic
9. horticulture
10. Fundy

Brazil

1. Andes
2. Easter
3. Aconcagua
4. Venezuela
5. Brasilia
6. Portuguese
7. Amazon
8. Capybaras or Capivaras
9. 1500s
10. Anacondas

Bolivia

1. Bravery and the blood of national heroes
2. La Paz
3. Boliviano
4. skunk
5. Uros
6. stones
7. 1552
8. rooster
9. adobe
10. five

Norway

1. Elbrus
2. Bishop
3. Vatican
4. Oslo
5. Norwegian krone
6. dawn
7. bunad
8. fjords
9. Lucia
10. king

Italy

1. Rome
2. rebirth
3. republic
4. French
5. Euro
6. Supper
7. slash
8. Disciple's
9. Constantine
10. 80

France

1. Paris
2. Lascaux
3. Gallic rooster
4. Clotilde
5. Chartres
6. Tour
7. cheeses
8. Romani
9. lavender
10. Eiffel

Germany

1. The Holy Roman Emperor
2. Berlin
3. Autobahn
4. Bible
5. by God's grace alone
6. *Soli Deo Gloria* (to God alone be glory)
7. Nutcracker
8. Swan Stone
9. cycle
10. Luther

United Kingdom

1. London
2. British pound
3. lion, unicorn, red dragon, and harp
4. St. Andrews
5. Wilberforce
6. 1611
7. puffins
8. *Canterbury*
9. Snowdonia
10. Giants

Russia

1. Moscow
2. Russian ruble
3. Trans-Siberian
4. God
5. Synodal
6. Peter
7. eleven
8. Nerpa or Baikal
9. tundra
10. Cyrillic

Africa

1. Victoria
2. Kilimanjaro
3. Kariba
4. Nile
5. Livingstone
6. Nairobi
7. Kenyan shilling
8. Victoria
9. pangolins
10. savannas

Egypt

1. Cairo
2. Arabic
3. golden
4. Kemet
5. Sahara
6. Bedouins
7. Giza
8. Muslims
9. Herod's
10. Coptic

South Africa

1. unity
2. Pretoria
3. Republic
4. diamonds
5. cross
6. Chacma
7. Setswana
8. Lovedale
9. Hope
10. Agulhas

Israel

1. 17
2. Everest
3. Petra
4. Europe
5. *bhalu*
6. Jerusalem
7. Dead
8. Peace
9. hoopoe
10. Kibbutz

Saudi Arabia

1. Riyadh
2. Monarchy
3. Saudi riyal
4. Mecca
5. hajj
6. three
7. calligraphy
8. Jubail
9. Islam
10. oryx

India

1. chakra
2. New Delhi
3. Indian rupee
4. Ganges or Ganga
5. Mahal
6. Teresa
7. Carey
8. Hindi
9. bond
10. snow

China

1. Beijing
2. Dragon
3. Grand
4. terra cotta
5. Tang
6. Taylor
7. Yangtze
8. alphabet
9. Silk
10. New

Japan

1. origin
2. Tokyo
3. archipelago
4. Ukiyo-e
5. Kakure
6. Bonsai
7. oragami
8. Fuji
9. Children's
10. serow

Malaysia

1. Kuala Lumpur
2. Constitutional monarchy
3. Tiger
4. 13
5. Petronas
6. Sarawak
7. rafflesia
8. Persian
9. Basel
10. sun

Australia

1. Tasmanian
2. Kosciuszko
3. Reefs
4. saltwater
5. Canberra
6. Sydney
7. Murray
8. Cook
9. fruit
10. didgeridoo

New Zealand

1. Cross
2. Wellington
3. Fjords
4. kiwi
5. Maori
6. Kiwifruit
7. Milford
8. Cook
9. Waiotapu
10. English

Antarctica

1. Lake Vostok
2. Emperor
3. 90
4. wingless
5. Vinson
6. driest
7. krill
8. Norwegian
9. Pole
10. Trinity

Children's Atlas of God's World ⟜ Quiz Answer Key

Biomes of the World

1. biological
2. shadow
3. snowfall
4. Grasslands
5. Deserts
6. poor
7. leaves
8. Barrier
9. Mountains
10. tundra

From wind and weather to constellations and comets!

The Investigate the Possibilities books incorporate inexpensive and easy experiments to teach important concepts for students grades 3 to 8! Learn about how weather events like Noah's Flood impact our world, and:

- Discover the amazing effects of weather around the globe
- Understand how the solar system and universe works
- Provides one year of science that includes a weekly lesson schedule, quizzes and tests, answer keys, and master supply list.

This parent lesson planner offers two levels of testing completing each set of books, and helping busy educators/parents teach multi-grade levels at one time!

Science Starters Package:

Water & Weather (Text, Student, & Teacher Books) & The Universe (Text, Student, & Teacher Books) & Parent Lesson Planner
ISBN: 978-0-89051-816-8 $54.99

Also Available in this Series:
Forces & Motion | Elementary Physical Science
The Earth | Elementary Earth Science
Matter | Chemistry
Energy | Physics
Teacher/student guides also available.

Master Books®
Curriculum

Now turn your favorite **Master Books** into curriculum! Each Parent Lesson Plan (PLP) includes:

- An easy-to-follow, one-year educational calendar
- Helpful worksheets, quizzes, tests, and answer keys
- Additional teaching helps and insights
- Complete with all you need to quickly and easily begin your education program today!

ELEMENTARY ZOOLOGY

1 year
4th – 6th

Package Includes: *World of Animals, Dinosaur Activity Book, The Complete Aquarium Adventure, The Complete Zoo Adventure, Parent Lesson Planner*

5 Book Package
978-0-89051-747-5 $84.99

SCIENCE STARTERS: ELEMENTARY PHYSICAL & EARTH SCIENCE

1 year
3rd – 8th grade

6 Book Package Includes: *Forces & Motion –Student, Student Journal, and Teacher; The Earth – Student, Teacher & Student Journal; Parent Lesson Planner*

6 Book Package
978-0-89051-748-2 $51.99

SCIENCE STARTERS: ELEMENTARY CHEMISTRY & PHYSICS

1 year
3rd – 8th grade

Package Includes: *Matter – Student, Student Journal, and Teacher; Energy – Student, Teacher, & Student Journal; Parent Lesson Planner*

7 Book Package
978-0-89051-749-9 $54.99

INTRO TO METEOROLOGY & ASTRONOMY

1 year
7th – 9th grade
½ Credit

Package Includes: *The Weather Book; The Astronomy Book; Parent Lesson Planner*

3 Book Package
978-0-89051-753-6 $44.99

INTRO TO OCEANOGRAPHY & ECOLOGY

1 year
7th – 9th grade
½ Credit

Package Includes: *The Ocean Book; The Ecology Book; Parent Lesson Planner*

3 Book Package
978-0-89051-754-3 $45.99

INTRO TO SPELEOLOGY & PALEONTOLOGY

1 year
7th – 9th grade
½ Credit

Package Includes: *The Cave Book; The Fossil Book; Parent Lesson Planner*

3 Book Package
978-0-89051-752-9 $44.99

CONCEPTS OF MEDICINE & BIOLOGY

1 year
7th – 9th grade
½ Credit

Package Includes: *Exploring the History of Medicine; Exploring the World of Biology; Parent Lesson Planner*

3 Book Package
978-0-89051-756-7 $40.99

CONCEPTS OF MATHEMATICS & PHYSICS

1 year
7th – 9th grade
½ Credit

Package Includes: *Exploring the World of Mathematics; Exploring the World of Physics; Parent Lesson Planner*

3 Book Package
978-0-89051-757-4 $40.99

CONCEPTS OF EARTH SCIENCE & CHEMISTRY

1 year
7th – 9th grade
½ Credit

Package Includes: *Exploring Planet Earth; Exploring the World of Chemistry; Parent Lesson Planner*

3 Book Package
978-0-89051-755-0 $40.99

THE SCIENCE OF LIFE: BIOLOGY

1 year
8th – 9th grade
½ Credit

Package Includes: *Building Blocks in Science; Building Blocks in Life Science; Parent Lesson Planner*

3 Book Package
978-0-89051-758-1 $44.99

BASIC PRE-MED

1 year
8th – 9th grade
½ Credit

Package Includes: *The Genesis of Germs; The Building Blocks in Life Science; Parent Lesson Planner*

3 Book Package
978-0-89051-759-8 $43.99

INTRO TO ASTRONOMY

1 year
7th – 9th grade
½ Credit

Package Includes: *The Stargazer's Guide to the Night Sky; Parent Lesson Planner*

2 Book Package
978-0-89051-760-4 $47.99

INTRO TO ARCHAEOLOGY & GEOLOGY

1 year
7th – 9th
½ Credit

Package Includes: *The Archaeology Book; The Geology Book; Parent Lesson Planner*

3 Book Package
978-0-89051-751-2 $45.99

SURVEY OF SCIENCE HISTORY & CONCEPTS

1 year
10th – 12th grade
1 Credit

Package Includes: *The World of Mathematics; The World of Physics; The World of Biology; The World of Chemistry; Parent Lesson Planner*

5 Book Package
978-0-89051-764-2 $72.99

SURVEY OF SCIENCE SPECIALTIES

1 year
10th – 12th grade
1 Credit

Package Includes: *The Cave Book; The Fossil Book; The Geology Book; The Archaeology Book; Parent Lesson Planner*

5 Book Package
978-0-89051-765-9 $81.99

SURVEY OF ASTRONOMY

1 year
10th – 12th grade
1 Credit

Package Includes: *The Stargazers Guide to the Night Sky; Taking Back Astronomy; Our Created Moon DVD; Created Cosmos DVD; Parent Lesson Planner*

4 Book, 2 DVD Package
978-0-89051-766-6 $113.99

GEOLOGY & BIBLICAL HISTORY

1 year
8th – 9th
1 Credit

Package Includes: *Explore the Grand Canyon; Explore Yellowstone; Explore Yosemite & Zion National Parks; Your Guide to the Grand Canyon; Your Guide to Yellowstone; Your Guide to Zion & Bryce Canyon National Parks; Parent Lesson Planner.*

4 Book, 3 DVD Package
978-0-89051-750-5 $108.99

PALEONTOLOGY: LIVING FOSSILS

1 year
10th – 12th grade
½ Credit

Package Includes: *Living Fossils, Living Fossils Teacher Guide, Living Fossils DVD; Parent Lesson Planner*

3 Book, 1 DVD Package
978-0-89051-763-5 $66.99

LIFE SCIENCE ORIGINS & SCIENTIFIC THEORY

1 year
10th – 12th grade
1 Credit

Package Includes: *Evolution: the Grand Experiment, Teacher Guide, DVD; Living Fossils, Teacher Guide, DVD; Parent Lesson Planner*

5 Book, 2 DVD Package
978-0-89051-761-1 $144.99

NATURAL SCIENCE THE STORY OF ORIGINS

1 year
10th – 12th grade
½ Credit

Package Includes: *Evolution: the Grand Experiment; Evolution: the Grand Experiment Teacher's Guide; Evolution: the Grand Experiment DVD; Parent Lesson Planner*

3 Book, 1 DVD Package
978-0-89051-762-8 $71.99

ADVANCED PRE-MED STUDIES

1 year
10th – 12th grade
1 Credit

Package Includes: *Building Blocks in Life Science; The Genesis of Germs; Body by Design; Exploring the History of Medicine; Parent Lesson Planner*

5 Book Package
978-0-89051-767-3 $76.99

BIBLICAL ARCHAEOLOGY

1 year
10th – 12th grade
1 Credit

Package Includes: *Unwrapping the Pharaohs; Unveiling the Kings of Israel; The Archaeology Book; Parent Lesson Planner*

4 Book Package
978-0-89051-768-0 $99.99

CHRISTIAN HERITAGE

1 year
10th – 12th grade
1 Credit

Package Includes: *For You They Signed; Lesson Parent Planner*

2 Book Package
978-0-89051-769-7 $50.99

SCIENCE STARTERS: ELEMENTARY GENERAL SCIENCE & ASTRONOMY

1 year
3rd – 8th grade

Package Includes: *Water & Weather – Student, Student Journal, and Teacher; The Universe – Student, Teacher, & Student Journal; Parent Lesson Planner*

7 Book Package
978-0-89051-816-8 $54.99

ELEMENTARY WORLD HISTORY

1 year
5th – 8th

Package Includes: *The Big Book of History; Noah's Ark: Thinking Outside the Box (book and DVD); & Parent Lesson Planner*

3 Book, 1 DVD Package
978-0-89051-815-1 $66.96

ELEMENTARY GEOGRAPHY AND CULTURES

1 year
3rd – 6th grade

Package Includes: *Children's Atlas of God's World, Passport to the World, & Parent Lesson Planner*

3 Book Package
978-0-89051-814-4 $49.99

APPLIED SCIENCE: STUDIES OF GOD'S DESIGN IN NATURE

1 year
7th – 9th grade
1 Credit

Package Includes: *Made in Heaven, Champions of Invention, Discovery of Design, & Parent Lesson Planner*

4 Book Package
978-0-89051-812-0 $50.99

CONCEPTS OF BIOGEOLOGY & ASTRONOMY

1 year
7th – 9th grade
½ Credit

Package Includes: *Exploring the World Around You, Exploring the World of Astronomy, & Parent Lesson Planner*

3 Book Package
978-0-89051-813-7 $41.99

INTRO TO BIBLICAL GREEK

½ year language studies
7th – 12th
½ Credit

Package Includes: *It's Not Greek to Me DVD & Parent Lesson Planner*

1 Book, 1 DVD Package
978-0-89051-818-2 $33.99

INTRO TO ECONOMICS: MONEY, HISTORY, & FISCAL FAITH

½ year economics
9th – 12th
½ Credit

Package Includes: *Bankruptcy of Our Nation, Money Wise DVD, & Parent Lesson Planner*

2 Book, 4 DVD Package
978-0-89051-811-3 $57.99

Master Books®

P.O. Box 726
Green Forest, AR 72638

Visit masterbooks.net for additional information, look insides, video trailers, and more!